Get Ready!

FOR STANDARDIZED TESTS

GRADE SIX

Other Books in the *Get Ready!* Series:

Get Ready! for Standardized Tests: Grade 1 by Joseph Harris, Ph.D.

Get Ready! for Standardized Tests: Grade 2 by Joseph Harris, Ph.D.

Get Ready! for Standardized Tests: Grade 3 by Karen Mersky, Ph.D.

Get Ready! for Standardized Tests: Grade 4 by Joseph Harris, Ph.D.

Get Ready! for Standardized Tests: Grade 5 by Leslie E. Talbott, Ph.D.

TEST PREPARATION SERIES

Get Ready!

FOR STANDARDIZED TESTS

GRADE SIX

Shirley Vickery, Ph.D.

Carol Turkington
Series Editor

McGraw-Hill

New York   San Francisco   Washington, D.C.   Auckland   Bogotá
Caracas   Lisbon   London   Madrid   Mexico City   Milan
Montreal   New Delhi   San Juan   Singapore
Sydney   Tokyo   Toronto

Library of Congress Cataloging-in-Publication Data

Get ready! for standardized tests / c Carol Turkington, series editor.
      p. cm.
    Includes bibliographical references.
    Contents: [1] Grade 1 / Joseph Harris — [2] Grade 2 / Joseph Harris — [3] Grade 3 / Karen Mersky — [4] Grade 4 / Joseph Harris — [5] Grade 5 / Leslie E. Talbott — [6] Grade 6 / Shirley Vickery.
    ISBN 0-07-136010-7 (v. 1) — ISBN 0-07-136011-5 (v. 2) — ISBN 0-07-136012-3 (v. 3) — ISBN 0-07-136013-1 (v. 4) — ISBN 0-07-136014-X (v. 5) — ISBN 0-07-136015-8 (v. 6)
    1. Achievement tests—United States—Study guides.   2. Education, Elementary—United States—Evaluation.   3. Education, Elementary—Parent participation—United States.   I. Turkington, Carol. II. Harris, Joseph.

LB3060.22 .G48   2000
372.126—dc21
                                                  00-056083

McGraw-Hill

*A Division of The **McGraw·Hill** Companies*

2 3 4 5 6 7 8 9 0 PBT/PBT 0 9 8 7 6 5 4 3 2 1 0

ISBN 0-07-136015-8

This book was set in New Century Schoolbook by Inkwell Publishing Services.

Printed and bound by Phoenix Book Technology.

McGraw-Hill books are available at special quantity discounts to use as premiums and sales promotions, or for use in corporate training programs. For more information, please write to the Director of Special Sales, McGraw-Hill, Professional Publishing, Two Penn Plaza, New York, NY 10121-2298. Or contact your local bookstore.

To Lex, Ellen, and Annie

Contents

Acknowledgments

I wish to acknowledge the contributions of multiple individuals to the success of this project. Special thanks go to the hundreds of teachers, staff, parents, and students of Richland School District Two in Columbia, South Carolina, with whom I have worked for the past twenty years. From this group, I borrowed and stole many ideas about educating children. Many of the suggestions in this book were garnered from conferences and observations in hundreds of exciting, focused classes. My own children, Ellen and Annie Rogerson, helped me learn what it is like for parents to follow up on an endless array of school projects, homework assignments, and after-school activities. I owe thanks to Joe Harris for suggesting that I participate in this project, and to Carol Turkington for providing editorial advice and encouragement. Finally, thanks to my husband, Lex Rogerson, who made library trips, suggested sample items, and provided unending moral support.

SKILLS CHECKLIST

MY CHILD ...	HAS LEARNED	IS WORKING ON
VOCABULARY		
Synonyms		
Antonyms		
Homophones		
Multi-meaning words		
Words in context		
Word study		
Word definitions		
READING COMPREHENSION		
Determining the main idea of a passage		
Noting patterns and trends		
Analyzing information critically		
Identifying details		
LANGUAGE MECHANICS		
Capitalization		
Punctuation		
LANGUAGE EXPRESSION		
Grammar		
SPELLING AND STUDY SKILLS		
Editing		
Word components		
Study skills		
Developing graphic organizers		
Developing an outline		
Using reference materials		
MATH CONCEPTS		
Numeration		
Number concepts		
Number properties		
Fractions		
Decimals		
MATH COMPUTATION		
Addition		
Subtraction		
Multiplication		
Division		
MATH APPLICATIONS		
Geometry		
Measurement		
Problem solving		
Data analysis		
Algebra		

Introduction

Almost all of us have taken standardized tests in school. We spent several days bubbling-in answers, shifting in our seats. No one ever told us why we took the tests or what they would do with the results. We just took them and never heard about them again.

Today many parents aren't aware they are entitled to see their children's permanent records and, at a reasonable cost, to obtain copies of any information not protected by copyright, including testing scores. Late in the school year, most parents receive standardized test results with confusing bar charts and detailed explanations of scores that few people seem to understand.

In response to a series of negative reports on the state of education in this country, Americans have begun to demand that something be done to improve our schools. We have come to expect higher levels of accountability as schools face the competing pressures of rising educational expectations and declining school budgets. High-stakes standardized tests are rapidly becoming the main tool of accountability for students, teachers, and school administrators. If students' test scores don't continually rise, teachers and principals face the potential loss of school funding and, ultimately, their jobs. Summer school and private after-school tutorial program enrollments are swelling with students who have not met score standards or who, everyone agrees, could score higher.

While there is a great deal of controversy about whether it is appropriate for schools to use standardized tests to make major decisions about individual students, it appears likely that standardized tests are here to stay. They will be used to evaluate students, teachers, and the schools; schools are sure to continue to use students' test scores to demonstrate their accountability to the community.

The purposes of this guide are to acquaint you with the types of standardized tests your children may take; to help you understand the test results; and to help you work with your children in skill areas that are measured by standardized tests so they can perform as well as possible.

Types of Standardized Tests

The two major types of group standardized tests are *criterion-referenced tests* and *norm-referenced tests*. Think back to when you learned to tie your shoes. First Mom or Dad showed you how to loosen the laces on your shoe so that you could insert your foot; then they showed you how to tighten the laces—but not too tight. They showed you how to make bows and how to tie a knot. All the steps we just described constitute what is called a *skills hierarchy:* a list of skills from easiest to most difficult that are related to some goal, such as tying a shoelace.

Criterion-referenced tests are designed to determine at what level students are performing on various skills hierarchies. These tests assume that development of skills follows a sequence of steps. For example, if you were teaching shoelace tying, the skills hierarchy might appear this way:

1. Loosen laces.
2. Insert foot.
3. Tighten laces.
4. Make loops with both lace ends.
5. Tie a square knot.

Criterion-referenced tests try to identify how far along the skills hierarchy the student has progressed. There is no comparison against anyone else's score, only against an expected skill level. The main question criterion-referenced tests ask is: "Where is this child in the development of this group of skills?"

Norm-referenced tests, in contrast, are typically constructed to compare children in their abilities as to different skills areas. Although the experts who design test items may be aware of skills hierarchies, they are more concerned with how much of some skill the child has mastered, rather than at what level on the skills hierarchy the child is.

Ideally, the questions on these tests range from very easy items to those that are impossibly difficult. The essential feature of norm-referenced tests is that scores on these measures can be compared to scores of children in similar groups. They answer this question: "How does the child compare with other children of the same age or grade placement in the development of this skill?"

This book provides strategies for increasing your child's scores on both standardized norm-referenced and criterion-referenced tests.

The Major Standardized Tests

Many criterion-referenced tests currently in use are created locally or (at best) on a state level,

and there are far too many of them to go into detail here about specific tests. However, children prepare for them in basically the same way they do for norm-referenced tests.

A very small pool of norm-referenced tests is used throughout the country, consisting primarily of the Big Five:

- California Achievement Tests (CTB/McGraw-Hill)
- Iowa Tests of Basic Skills (Riverside)
- Metropolitan Achievement Test (Harcourt-Brace & Company)
- Stanford Achievement Test (Psychological Corporation)
- TerraNova [formerly Comprehensive Test of Basic Skills] (McGraw-Hill)

These tests use various terms for the academic skills areas they assess, but they generally test several types of reading, language, and mathematics skills, along with social studies and science. They may include additional assessments, such as of study and reference skills.

How States Use Standardized Tests

Despite widespread belief and practice to the contrary, group standardized tests are designed to assess and compare the achievement of groups. They are *not* designed to provide detailed diagnostic assessments of individual students. (For detailed individual assessments, children should be given individual diagnostic tests by properly qualified professionals, including trained guidance counselors, speech and language therapists, and school psychologists.) Here are examples of the types of questions group standardized tests are designed to answer:

- How did the reading achievement of students at Valley Elementary School this year compare with their reading achievement last year?

- How did math scores at Wonderland Middle School compare with those of students at Parkside Middle School this year?

- As a group, how did Hilltop High School students compare with the national averages in the achievement areas tested?

- How did the district's first graders' math scores compare with the district's fifth graders' math scores?

The fact that these tests are designed primarily to test and compare groups doesn't mean that test data on individual students isn't useful. It does mean that when we use these tests to diagnose individual students, we are using them for a purpose for which they were not designed.

Think of group standardized tests as being similar to health fairs at the local mall. Rather than check into your local hospital and spend thousands of dollars on full, individual tests for a wide range of conditions, you can go from station to station and take part in different health screenings. Of course, one would never diagnose heart disease or cancer on the basis of the screening done at the mall. At most, suspicious results on the screening would suggest that you need to visit a doctor for a more complete examination.

In the same way, group standardized tests provide a way of screening the achievement of many students quickly. Although you shouldn't diagnose learning problems solely based on the results of these tests, the results can tell you that you should think about referring a child for a more definitive, individual assessment.

An individual student's group test data should be considered only a point of information. Teachers and school administrators may use standardized test results to support or question hypotheses they have made about students; but these scores must be used alongside other information, such as teacher comments, daily work, homework, class test grades, parent observations, medical needs, and social history.

Valid Uses of Standardized Test Scores

Here are examples of appropriate uses of test scores for individual students:

- Mr. Cone thinks that Samantha, a third grader, is struggling in math. He reviews her file and finds that her first- and second-grade standardized test math scores were very low. Her first- and second-grade teachers recall episodes in which Samantha cried because she couldn't understand certain math concepts, and mention that she was teased by other children, who called her "Dummy." Mr. Cone decides to refer Samantha to the school assistance team to determine whether she should be referred for individual testing for a learning disability related to math.

- The local college wants to set up a tutoring program for elementary school children who are struggling academically. In deciding which youngsters to nominate for the program, the teachers consider the students' averages in different subjects, the degree to which students seem to be struggling, parents' reports, and standardized test scores.

- For the second year in a row, Gene has performed poorly on the latest round of standardized tests. His teachers all agree that Gene seems to have some serious learning problems. They had hoped that Gene was immature for his class and that he would do better this year; but his dismal grades continue. Gene is referred to the school assistance team to determine whether he should be sent to the school psychologist for assessment of a possible learning handicap.

Inappropriate Use of Standardized Test Scores

Here are examples of how schools have sometimes used standardized test results inappropriately:

- Mr. Johnson groups his students into reading groups solely on the basis of their standardized test scores.

- Ms. Henry recommends that Susie be held back a year because she performed poorly on the standardized tests, despite strong grades on daily assignments, homework, and class tests.

- Gerald's teacher refers him for consideration in the district's gifted program, which accepts students using a combination of intelligence test scores, achievement test scores, and teacher recommendations. Gerald's intelligence test scores were very high. Unfortunately, he had a bad cold during the week of the standardized group achievement tests and was taking powerful antihistamines, which made him feel sleepy. As a result, he scored too low on the achievement tests to qualify.

The public has come to demand increasingly high levels of accountability for public schools. We demand that schools test so that we have hard data with which to hold the schools accountable. But too often, politicians and the public place more faith in the test results than is justified. Regardless of whether it's appropriate to do so and regardless of the reasons schools use standardized test results as they do, many schools base crucial programming and eligibility decisions on scores from group standardized tests. It's to your child's advantage, then, to perform as well as possible on these tests.

Two Basic Assumptions

The strategies we present in this book come from two basic assumptions:

1. Most students can raise their standardized test scores.

2. Parents can help their children become stronger in the skills the tests assess.

This book provides the information you need to learn what skill areas the tests measure, what general skills your child is being taught in a particular grade, how to prepare your child to take the tests, and what to do with the results. In the appendices you will find information to help you decipher test interpretations; a listing of which states currently require what tests; and additional resources to help you help your child to do better in school and to prepare for the tests.

A Word about Coaching

This guide is *not* about coaching your child. When we use the term *coaching* in referring to standardized testing, we mean trying to give someone an unfair advantage, either by revealing beforehand what exact items will be on the test or by teaching "tricks" that will supposedly allow a student to take advantage of some detail in how the tests are constructed.

Some people try to coach students in shrewd test-taking strategies that take advantage of how the tests are supposedly constructed rather than strengthening the students' skills in the areas tested. Over the years, for example, many rumors have been floated about "secret formulas" that test companies use.

This type of coaching emphasizes ways to help students obtain scores they didn't earn—to get something for nothing. Stories have appeared in the press about teachers who have coached their students on specific questions, parents who have tried to obtain advance copies of tests, and students who have written down test questions after taking standardized tests and sold them to others. Because of the importance of test security, test companies and states aggressively prosecute those who attempt to violate test security—and they should do so.

How to Raise Test Scores

Factors that are unrelated to how strong students are but that might artificially lower test scores include anything that prevents students from making scores that accurately describe their actual abilities. Some of those factors are:

- giving the tests in uncomfortably cold or hot rooms;
- allowing outside noises to interfere with test taking; and
- reproducing test booklets in such small print or with such faint ink that students can't read the questions.

Such problems require administrative attention from both the test publishers, who must make sure that they obtain their norms for the tests under the same conditions students face when they take the tests; and school administrators, who must ensure that conditions under which their students take the tests are as close as possible to those specified by the test publishers.

Individual students also face problems that can artificially lower their test scores, and parents can do something about many of these problems. Stomach aches, headaches, sleep deprivation, colds and flu, and emotional upsets due to a recent tragedy are problems that might call for the student to take the tests during make-up sessions. Some students have physical conditions such as muscle-control problems, palsies, or difficulty paying attention that require work over many months or even years before students can obtain accurate test scores on standardized tests. And, of course, some students just don't take the testing seriously or may even intentionally perform poorly. Parents can help their children overcome many of these obstacles to obtaining accurate scores.

Finally, with this book parents are able to help their children raise their scores by:

- increasing their familiarity (and their comfort level) with the types of questions on standardized tests;
- drills and practice exercises to increase their skill in handling the kinds of questions they will meet; and

- providing lots of fun ways for parents to help their children work on the skill areas that will be tested.

Test Questions

The favorite type of question for standardized tests is the multiple-choice question. For example:

1. The first President of the United States was:

   A Abraham Lincoln

   B Martin Luther King, Jr.

   C George Washington

   D Thomas Jefferson

The main advantage of multiple-choice questions is that it is easy to score them quickly and accurately. They lend themselves to optical scanning test forms, on which students fill in bubbles or squares and the forms are scored by machine. Increasingly, companies are moving from paper-based testing to computer-based testing, using multiple-choice questions.

The main disadvantage of multiple-choice questions is that they restrict test items to those that can be put in that form. Many educators and civil rights advocates have noted that the multiple-choice format only reveals a superficial understanding of the subject. It's not possible with multiple-choice questions to test a student's ability to construct a detailed, logical argument on some issue or to explain a detailed process. Although some of the major tests are beginning to incorporate more subjectively scored items, such as short answer or essay questions, the vast majority of test items continue to be in multiple-choice format.

In the past, some people believed there were special formulas or tricks to help test-takers determine which multiple-choice answer was the correct one. There may have been some truth to *some* claims for past tests. Computer analyses of some past tests revealed certain biases in how tests were constructed. For example, the old advice to pick *D* when in doubt appears to have been valid for some past tests. However, test publishers have become so sophisticated in their ability to detect patterns of bias in the formulation of test questions and answers that they now guard against it aggressively.

In Chapter 1, we provide information about general test-taking considerations, with advice on how parents can help students overcome testing obstacles. The rest of the book provides information to help parents help their children strengthen skills in the tested areas.

Joseph Harris, Ph.D.

Test-Taking Basics

Parents everywhere are hearing about the importance of standardized testing to their students' academic performance and record. Sixth grade is a year when the testing demands for students seem to increase. Many states don't test younger students, and others only begin to use the results for serious purposes when students get to sixth grade. Test scores are used in a variety of ways in different districts, but many schools use tests to determine the need for more tutoring or for summer school, or to determine promotion to the next grade. They also may be used to identify students for gifted and talented programs. As a result, the standardized testing process may provoke anxiety for students and parents alike.

We hope that this chapter will ease some of that anxiety by giving students and parents an idea of what to expect and how to prepare for the testing process in general.

How Parents Can Help

Where do you start if you want to work with your sixth grader to prepare for standardized tests? In general, you should remember that your child may be turned off or unmotivated by testing.

Students who are weak in academic skills may be very nervous and insecure about the testing process and may resist the efforts of their parents to help them. Parents who are frustrated with the process and with their children may only compound the problems by increasing the child's frustration level with both school and testing.

It's important that resistance doesn't become a habit for your child. In order to avoid this, try to remember:

- You can't improve test scores overnight.

- Test scores are the result of work done all year long, not just the few weeks before the test.

- Students do best on tests when they are somewhat concerned and motivated, but not so anxious that they have trouble concentrating.

- A student's level of anxiety about a situation is usually directly related to your own concern.

- The work assigned in school should directly relate to skills tested on the end-of-the-year assessment. Therefore, students should place a high priority on doing assigned homework and schoolwork, following the instruction in class, studying for tests, and reviewing assignments.

- Students who do well in their daily work usually do well in standardized testing.

- Some students who do well in daily work don't do well in standardized testing. This often frustrates parents and students alike. These students may very well be helped by the suggestions found in this book.

Tests are different from daily assignments in that they present different skills on the same page. Skills are intermixed on the page and students are expected to shift from skill to skill without warning. In class, students may receive practice on one or a few skills at a time. It's much harder to take a test when items of different kinds are mixed on the same page.

It's important to prepare your child to consider that he will encounter some items on standardized tests that he's never seen before and hasn't been taught. This practice of providing some "ceiling" or upper limit for the test actually allows the test to measure the limits of a child's knowledge. While some students may see this as a frustrating part of the test, all tests of this nature have some items almost no one gets right. It's a predictable part of the process. Students who realize that they won't know the answers to all questions usually feel better about the testing process in general.

The Sixth Grader's Development

Sixth graders are typically between 11 and 12 years old and at some stage of beginning adolescence. This is an exciting time of great development—physically, socially, and mentally. Parents are often frustrated with the changes they see in their adolescents, and it's easy to understand why. In order to understand the implications these changes have for your child's thinking skills, you need to start thinking about your child in different ways, and to work with him as a budding adult rather than as a child. That is no small task!

Let's look at a few of the important characteristics of a sixth grader's thinking and how they affect school and standardized testing.

Abstract Reasoning Skills

Sixth graders start relying more on their ability to use abstract reasoning skills. Someone who thinks abstractly looks beyond the information right in front of him and reasons on a wider level. Most of the time, adults rely mostly on concrete reasoning skills: We turn on a TV or start a car and we don't think about how they work. If a machine doesn't work when we try to

turn it on, we may try a second time, but eventually we call a repairperson rather than try to figure it out ourselves. When we think abstractly, we think about the way the machine works and make guesses about how to repair it. We develop explanations for the problem.

Sixth graders are just beginning to develop this abstract ability to take themselves away from the specifics of a situation and guess what the situation means. Sixth-grade teachers can often capture the attention of their students by asking questions like:

"What does this mean?"

"How does this work?"

"How can this be done?"

The ability to reason on a higher level will eventually lead sixth graders to wonder about more complicated questions that have no easy answers. Questions of a social, ethical, or religious nature may occupy some adolescents in very intense ways. They may spend hours pondering who they are, when parents and teachers may wish they would get busy with assigned work!

Creativity

Sixth graders may think very creatively and have boundless energy for composing, writing, designing, or drawing. The products may not be quite so refined as they will be in a few more years, but the sheer joy of developing and planning may be exhilarating to children this age. This creativity can lead to beautiful poetry, music, or stories that contain hints of the complex thinking adolescents are beginning to do.

What happens to creativity at home and at school? You should be aware of your child's need to express his boundless creativity and try to find ways to channel it into learning activities. If creativity is quashed, the student may become less motivated and may begin to see school as a very boring place where no new ideas or suggestions are tolerated. Students may develop similar feelings about their parents' help.

How You Can Help Your Child

You can help to boost your child's motivation and provide outlets for that creativity in many ways:

- Praise your child's writing and allow him to share it or not as he wishes.

- Listen to ideas and relate them to other information you've learned or discussed.

- Relate specific academic material covered in school to your child's ideas.

- Help your child find special classes in art, music, or composition; you may need to look outside school for some classes.

- Allow your child to explore different methods of expression without expecting perfection.

- Listen to your child's favorite music and discuss with him what he likes about it.

Need for Autonomy. Your sixth grader may begin to express opinions that are different from yours and will often question your expectations and limits. This certainly impacts school performance, since your child may decide he wants to make the decisions about what he learns and how he spends his time. Your child may be less willing to respond to schedules and structures you select.

This can make preparing for standardized tests particularly difficult. Your child may resist your efforts to improve his school performance because he sees this as his responsibility. Your child may especially dislike having you decide which topics he needs to review or practice.

For these reasons, it's most helpful if you can avoid focusing on specific skill reviews and drills. That's why we won't try to teach you how to conduct educational drills with your child.

Instead, we'll point out ways you can prepare your child for schoolwork and testing by noticing the academic nature of various activities and by encouraging students to solve problems. When you help your child learn to solve problems dealing with everyday events, he will see

the need for academic skills and will be more likely to master them. Remember, when students practice in the real world, they remember the skills much better than when all practice is in textbooks.

This idea is so important that it deserves a few minutes of focus before we move on. Parents can do a number of things with their children—everything from buying and cooking food to participating in Scouts and athletics—to provide important opportunities for academic skill development. These activities are examples of the kinds of problems your child will face in the future, and offer a host of possibilities for work in math and reading. The list includes:

- shopping

- cooking

- planning a vacation

- budgeting

- repairing a toilet

- planting shrubs

- growing vegetables

- deciding which new car to buy

Your sixth grader will be more motivated to work with you on testing material if he sees it as interesting or social. Here are some tips to make these activities fun:

Go shopping. Several sixth grade students may enjoy a shopping trip. Give your child a budget and ask that he and his friends make suggestions on how to spend the money by focusing on finding the best buys on clothing.

Plan party time. Don't just throw your child a party. Involve him! Turn it into a lesson in management and math by having your sixth grader and his friends plan the number of items needed, the food to be served, the items to be bought, and the budget for the party.

Use sports. A group of sixth graders can follow a sports team together and compare statistics as the team progresses through the season.

Go camping. If your child likes camping, have him plan what supplies are needed, shop for the items, locate appropriate recipes, and adjust the recipes for the number of people attending the trip.

Obviously, the list could go on and on. While these things may not seem like traditional "schoolwork," in fact students need to practice what they learn in a meaningful context or in some real-life situation. Your child will get the most out of any learning situation when he is doing something related to the learning, not just reading a book or participating in a discussion.

Finally, your child will learn better when he practices for small amounts at a time over several days, rather than cramming for hours the night before a test.

The Social Side of Thinking. It may not seem so, but your child's social life does have an effect on his thinking ability. Any sixth-grade teacher will tell you that the most important thing to a child of this age is his social life. Sixth graders are starting to become extremely attached to and influenced by their peers. They're interested in what other students their age are doing, and they are motivated by things of interest to other students and to themselves.

In some cases, students may become self-conscious about "being smart" and may not want to perform well on tests. Some sixth graders may not like to call attention to themselves as "the brain." If this is happening to your child, you'll have to work hard to overcome this feeling. Here's how:

- Ask teachers to offer incentives for work well done.
- Develop an incentive system for your child featuring music, free time, or time with friends.
- Develop a sense of self-confidence in your child.

- Encourage your child to invite friends over for study time.
- Plan a fun outing that involves some of the skill areas your child needs help with, and allow your child to invite a friend along.
- Check out videos that develop some of the same interests and skills, and invite friends to watch, too.

What the Tests May Ask

Tests for sixth graders are usually somewhat different from tests for younger children. They're not just harder, but the style and format of the tests may be different as well. These differences include smaller print, the use of embedded instructions, longer time limits, larger written passages, increased abstract thinking, and more open-ended questions.

Smaller Print

Sixth graders are accustomed to smaller print in textbooks and lessons, so the smaller print on the tests shouldn't be a problem for anyone except those with poor vision. Students should be certain to bring glasses or contacts with them to the testing session. Be sure to obtain correct lenses for your child long before testing day, so he can become used to the lenses before the testing period begins.

Longer Time Limits

Sixth graders are expected to work for longer periods of time without receiving directions or breaks. Your child may routinely work without stopping for 45 minutes or longer. Those who typically need lots of breaks to complete their work may find this aspect of testing somewhat difficult and should begin to prepare themselves for longer work periods without help. If your child has a problem with longer work periods, have him wear a wristwatch to remind him how much time is left. He can also practice working on homework for longer periods without asking for help or taking a break.

Embedded Instructions

Instructions for many tests are given at the beginning of the session. In tests for younger students, these instructions apply to the entire session until the teacher gives another set of instructions.

However, in sixth grade, tests often have new directions that the student is expected to read independently and to apply to the next set of questions. An adult won't intervene to help him understand. If your child is used to having an adult talk him through directions, he may have trouble with this independent format.

To prepare for this, have your child practice reading directions on homework by himself before asking for help. He should also reread the directions, think them through step by step, and pick out key words. This skill is very important and is discussed in more detail.

Longer Written Passages

Your sixth grader will be expected to have a longer attention span for reading and writing. He'll have to read several pages at a time without taking a break. With smaller print and the lack of pictures, the amount of reading may intimidate many students.

It's important to remember that your child will be expected to review the passage to find answers. You can help your child deal with this situation by teaching him to scan the material and questions before reading, and to focus on reading one paragraph at a time.

More Questions Requiring Abstract Reasoning

Because sixth graders are better at abstract reasoning than younger students, test makers add questions requiring students to infer meanings, draw conclusions, apply data to another situation, or make predictions about what may happen in the future.

Open-Ended (Not Multiple-Choice) Questions

This is an important new development in testing and one that is becoming increasingly prevalent on newer tests. It's hard to prepare fully for this format because it's relatively new, but you and your child should be aware of the format and begin practicing. These kinds of questions will be encountered in real-life academic tasks (and real-life examples don't always come in multiple-choices!).

Test-Taking Skills Your Child Needs

You can help your sixth grader improve his test-taking skills by helping him learn to follow directions, pay attention for longer periods of time, fill in an answer sheet correctly, use the information available, budget time, and develop a system of checking answers.

Following Directions

All elementary school students need to develop skills at following directions in order to improve test performance. Sixth graders need to pay particular attention to this area because it's likely there will be some instructions embedded in the test. This means that at the beginning of a test section, the teacher will begin timing the test. After students have worked through a section of the test, they will find a new set of directions that requires them to answer questions according to a different format. This style of testing requires careful attention to the material and close reading of the new instructions.

Knowing the meaning of direction words is also important to sixth graders, because they will likely hear longer, more standardized instructions than they did at an earlier age. They will probably encounter the following direction words:

Read

Mark

Bubble-in

Calculate

Identify

Consider

Draw conclusions

Compare

Contrast

While most of these words are familiar to your child, it's easy to imagine that he'll need practice on some words before he can use them effectively. Whenever possible, use these words every day and explain the meanings.

Paying Attention

Sixth graders are usually able to work for long periods of time without stopping, but they may not be used to doing it very often. During school and while doing homework, your child may stop working periodically and take regular breaks. This won't be possible during standardized tests.

Try to establish periods of time during which your child must work without interruption. Find a place without distractions and set a reasonable time period (such as 30 minutes) for your child to work. Set a timer to go off after the allotted period, and then have your child take a short break. As the year progresses, increase the length of the study periods to 45 or 55 minutes.

By the time standardized testing arrives, your child should be used to working for up to an hour without a break, and won't be surprised or intimidated by the prospect of a long work period.

Using an Answer Sheet

While most sixth graders have encountered answer sheets before, many students still have trouble filling in circles or bubbles. Answer sheets are usually scored by computer, so it's important that all bubbles be filled in correctly and that no stray marks appear on the paper. Most important, your child needs to mark the answer in the correct spot. It's very easy to lose your place on an answer sheet, marking two answers on one line or skipping a line. When this happens, your child will miss most of the questions following the one in which he lost his place. Make sure your child understands the following tips:

- Fill in the bubble completely.
- Don't make stray marks on the page.
- Stay on the correct line.
- Use a ruler if necessary to keep on the correct line.
- Check periodically to make sure you're marking the line that goes with the question you're answering.

Using Available Information

Some questions will contain longer explanations of material that the student is expected to read. Students will be expected to use the information in the question efficiently in order to answer the question.

You can help your child get ready to handle this type of question by leaving your child notes about chores for the day and praising him for following directions. When you write such notes, use direction words and explain anything your child doesn't understand.

Here's an example:

Today I need you to gather your clothes together and put them in the laundry basket. After you've finished this, collect all the recyclables and sort them into categories. Please write me a note to let me know how full the containers are so we can plan a trip to the recycling station. Then replace your winter clothes with your spring ones and stack the winter ones in a box.

It may be difficult to get your child to follow all of these directions, but you can set the stage and hope for the best! At least you will know you're teaching the right direction words, even if the chores don't get done.

Budgeting Time

Sixth graders are often poor judges of time and may use too much time on one question while not leaving enough time for the others. One important skill your child should develop is the ability to guess elapsed time and to budget the time he has to accomplish something. You can help your child do this by:

- discussing time needed for daily activities, like driving to school;

- buying your child a watch so he can keep track of the time as he works;

- asking your child to estimate how much time it will take to do something and then comparing his guess to the actual time elapsed;

- using a kitchen timer to time homework periods.

Checking Responses

Developing a system for checking responses is a helpful strategy for students who are taking standardized tests. Often students find they have omitted items, not filled in blanks completely, moved to an incorrect line to bubble-in answers, or just marked the wrong answer.

Students won't usually have time to recheck the entire test, but this shouldn't be necessary. There is usually enough time, however, to make sure they have answered all the questions, and to recheck answers they weren't sure about.

Here's a good strategy for checking work on a test:

- Glance over the answer sheet to make sure there is an answer for each item.

- Check that no item has two answers.

- Be sure that when you finished the test, you were on the right number for the question you answered.

- Make sure you have left no stray marks on the answer sheet.

- If you're allowed to use scratch paper, make a list of questions you weren't sure about. If scratch paper is forbidden but you can mark on the test, place a small check in the margin next to the items you're not sure about. If you can't do either of these, put a small pencil mark on the answer sheet next to the items you weren't sure about. Come back to these questions at the end of the test if you still have time.

With these observations and strategies, you are ready to begin to look at the test topics themselves and to think about ways you and your child can work on the areas that will be included on standardized tests.

CHAPTER 2

Vocabulary

It's hard to imagine a more important skill for students and adults than vocabulary. Most educators will tell you that students with good vocabularies are those who do well in school in general. A good vocabulary is a key to understanding written and spoken information.

What Your Sixth Grader Should Be Learning

Sixth graders are expected to understand many words and to make guesses about word meanings based on their knowledge of base words, word endings and beginnings, and context cues.

What Tests May Ask

Standardized tests for the sixth grade include a number of multiple-choice questions on the following vocabulary items:

- Synonyms
- Antonyms
- Homophones
- Multi-meaning words
- Words in context
- Word study
- Definitions

What You and Your Child Can Do

You can boost your sixth grader's vocabulary at home, without having to resort to drills and flash cards. Here are a few ways:

Read. If you're hoping to increase your child's vocabulary, remember that the best way to do so is to read, read, read. Readers encounter new words every time they open a book and use the information they read to figure out the word's meaning. Encountering the word once may make your child aware of that word, but encountering it several times through reading will boost the chances that it will be remembered.

In order to increase vocabulary, it's not necessary for your sixth grader to read the selections herself. It's just as helpful for a student to hear selections read out loud—and in some cases, it may be more helpful than having her try to read it herself. If a student doesn't properly sound out a target word, she won't remember it from speaking and she won't use prior knowledge to help her determine the meaning. On the other hand, when she hears you read the material and pronounce the word correctly, she'll often recognize the word. Even listening to books on tape is appropriate for sixth graders. Some students enjoy the same books adults enjoy, if the content is appropriate.

Talk! Speaking and listening are other ways of increasing vocabulary. Let's face it: Most parents don't have time to develop separate vocabulary lessons for their children. Parents are forced to use whatever time is available to them to help their children in school. For most of us, that means talking either over the dinner table or while riding in the car. Both of these times present great opportunities to talk and to increase exposure to words and word meanings.

After-School Activities. Students who participate in a wide variety of activities after school also tend to have good vocabularies. Learning words associated with fun activities is a great way to increase vocabulary. Words learned in context and practiced daily tend to be remembered. Those who play soccer, for example, learn the vocabulary associated with the game; those who dance ballet learn a totally different set of words. Consider this list of words associated with gymnastics:

routine

compulsory

aerial

scale

optional

extension

walkover

handspring

waltz step

grapevine

Musicians learn another set of words, including:

scale

note

andante

allegro

The list goes on and on. Just imagine how much more it means to read a story about dogs if you have learned the "language of dogs" while having the experience of choosing and caring for a dog of your own.

TIPS FOR TAKING VOCABULARY TESTS

- Multiple-choice items on vocabulary subtests usually have at least four choices. Two of the choices will most likely be clearly wrong, and the last two will be somewhat close. If you aren't sure of the answer, it's helpful to try to eliminate the two clearly wrong choices first. Then take a look at the other two choices and determine whether you know which is correct. If you can't tell, you can at least guess with some increased chance of guessing the correct answer, because you only have two choices left.
- Vocabulary sections on a standardized test won't always give you a sample sentence, but they'll often give you a phrase using the target word. It may help to substitute the answer choices in place of the target word to see if the choices sound like possibilities. If one sounds clearly incorrect, then you can eliminate that choice and concentrate on the others.
- Read all of the choices carefully before selecting your answer.
- Some vocabulary subtests will give you a sentence or paragraph with a selected word and ask you to choose a definition or synonym. If you encounter items like this, consider all of the choices for definitions or synonyms before making your choice.
- Other questions will leave a blank and ask you to choose a word that best goes in the blank. Try all of the choices in the blank before making your selection.
- Use your knowledge of basic word parts to help you find the meaning of words. If you've studied word roots or stems, this should help you tremendously. If not, you can still use your knowledge of word parts you have studied, such as *re-*, *un-*, *bio-*, and *hydro-*, to help you decide which word fits.

Synonyms

Sixth graders may encounter test items that ask them to identify synonyms, or words that mean the same or nearly the same. You can help your child learn how to find synonyms by trying some of these strategies:

Imitate. When speaking or listening to what others say, restate what they have said using a synonym.

RICK: You know, yesterday I had a <u>discussion</u> with Larry that I've been thinking about a lot.

DAD: Which <u>conversation</u> do you mean?

RICK: It was the one early in the day when he mentioned my playing on Saturday. I was really <u>irritated</u> by what he said.

DAD: I didn't realize it <u>annoyed</u> you so much.

Note that the underlined words in the first two lines have nearly the same meaning, and the underlined words in the last two lines also have nearly the same meaning. The father has merely restated what his son said using a different word. He has not only shown his son that he's listening to his conversation, but he's enhanced Rick's understanding of words at the same time.

Use a Thesaurus. Provide your child with a thesaurus to use when doing her homework or school reports.

Use a Variety of Words. When your child writes compositions, encourage her to use a variety of words to give her work more interest and precision. A student's initial try at a composition might sound something like this:

Today we worked on writing good introductions to our papers. When writing a paper, it is important to have a good introduction. An introduction tells the reader what the paper is going to be about and gets his attention. An introduction may give a little information about the topic of your paper but it does not go into detail.

Suggest that she rewrite the essay using synonyms to give more interest. The student might write:

Today we worked on writing good introductions to our compositions. When preparing a paper, it is important to have a good beginning. An introduction tells the reader what the paper is going to be about and gets his

attention. An introduction may give a little information about the topic of your essay but it does not go into detail.

Note that the words *composition* and *essay* have been substituted for the word *paper* and that the word *beginning* has been used instead of *introduction*. These replacements illustrate the use of synonyms and add interest for the reader.

Practice Skill: Synonyms

Directions: Look carefully at the answer choices. Choose the best synonym for the underlined word.

Example:

a <u>gorgeous</u> dress

- (A) ugly
- (B) beautiful
- (C) tiny
- (D) green

Answer:

- (B) beautiful

---

1 a <u>peculiar</u> story
- (A) ordinary
- (B) strange
- (C) specific
- (D) interesting

---

2 an <u>infinite</u> number
- (A) definite
- (B) negative
- (C) limitless
- (D) large

---

3 <u>sterilize</u> the skin
- (A) disinfect
- (B) cover
- (C) heal
- (D) contaminate

---

4 <u>cringing</u> near the door

   (A) standing

   (B) crying

   (C) crawling

   (D) cowering

(See page 113 for answer key.)

Antonyms

Antonyms are words meaning the opposite or nearly opposite of each other, such as create—destroy, cruel—kind, helpful—hurtful, or friendly–reserved.

What You and Your Child Can Do

Understanding a variety of antonyms also helps add variety to a student's writing and speaking. Not every word has an antonym, so it is sometimes a challenge to come up with examples. Parents can help students develop a greater sense of antonyms by trying some of these activities:

- Challenge your student to compete with you or someone else to see how many antonym combinations he can generate while you're in the car on the way to practice or school.
- Have someone describe an object using words meaning the opposite of what he or she really wants to say. Have other family members attempt to figure out what the individual is describing. Using the opposites game, a student might describe a tree as a short, indoor, nonliving thing. A building might be described as a living, soft, uninhabited thing.
- Practice describing differences between life today and life in the past using opposites. Include references to how people secured and cooked food and how they built houses and lived in them.
- Point out examples of antonyms in everyday conversation.

Practice Skill: Antonyms

Directions: Read each item. Choose the answer that means the opposite of the underlined word.

Example:

   the <u>lively</u> TV show

   (A) interesting

   (B) historical

   (C) dull

   (D) entertaining

Answer:

   (C) dull

5 the <u>sensational</u> news story

   (A) trivial

   (B) ordinary

   (C) outstanding

   (D) sensible

6 a <u>glum</u> family

   (A) concerned

   (B) intelligent

   (C) joyful

   (D) attractive

7 living in <u>bondage</u>

   (A) slavery

   (B) bandages

   (C) clusters

   (D) freedom

8 <u>clarifying</u> her position

   (A) confusing

   (B) describing

   (C) discussing

   (D) coaxing

(See page 113 for answer key.)

Homophones

Sixth graders may also encounter homophones—those tricky little words that sound the same but have different meanings. Examples include *to, two,* and *too; their, there,* and *they're;* and *red* and *read.* These words confuse even the best of students, and many adults! There isn't a good way to remember the differences, and there isn't usually a clue in the word or passage to let you know which spelling is correct in that situation. You just have to learn the words by practice.

What You and Your Child Can Do

There are a number of ways to practice the correct use of homophones:

- Notice homophones when you see them in your child's writing. Comment on them every time you see them. Notice it especially if your child has used the correct one!

- Discuss homophones as you encounter them in your writing. You may have to think before you write *there* or *their*, and it won't hurt for your child to know that.

- Make a list of homophones and place it in a convenient location. Add to the list as the week goes on. Reward your child for identifying a homophone not on the list.

Practice Skill: Homophones

Directions: Read each item. Fill in the blank with the correct answer.

Example:

I hope it doesn't _____.

  Ⓐ rein
  Ⓑ rain /
  Ⓒ reign
  Ⓓ none of the above

Answer:

  Ⓑ rain

9 I want to go _____.

  Ⓐ to
  Ⓑ too ⁄
  Ⓒ two
  Ⓓ none of the above

10 The girls liked to play music with _____ stereos.

  Ⓐ there
  Ⓑ they're
  Ⓒ their /
  Ⓓ none of the above

11 The boy put the saddle on his _____.

  Ⓐ burro
  Ⓑ burrow /
  Ⓒ borough
  Ⓓ none of the above

(See page 113 for answer key.)

Multi-Meaning Words

Multi-meaning words are two words that are spelled the same way but have different meanings. These words can be very confusing to students. Often, the meanings can be very different and can change the intent of a whole sentence. For example, *spring* can refer to a stream of water or a quick, forward motion. *General* can be an army officer or a lack of specificity.

What You and Your Child Can Do

- Challenge your student to come up with as many meanings of a common word as possible.

- While you're driving to soccer practice, make a list of several multi-meaning words and allow a certain amount of time for each family member to generate as many different meanings as possible.

- Challenge your child to develop a list of multi-meaning words.

- Keep a list of multi-meaning words on the refrigerator and watch how fast the list grows.

Practice Skill: Multi-Meaning Words

Directions: Choose the sentence in which the underlined word means the same thing as in the sentence given.

Example:

Where did I put my <u>comb</u>?

- Ⓐ The rooster shook his red <u>comb.</u>
- Ⓑ The boy dragged a <u>comb</u> through his red curls.
- Ⓒ I'll have to <u>comb</u> through these reports to find the answer.
- Ⓓ The bees stored their honey in a <u>comb.</u>

Answer:

- Ⓑ The boy dragged a <u>comb</u> through his red curls.

---

12 The trash was in a <u>compact</u> container and was easy to put in the dumpster.

- Ⓐ The nations made a <u>compact</u> to support each other in times of conflict.
- Ⓑ The <u>compact</u> made a quick left turn in front of the truck.
- Ⓒ She pulled her <u>compact</u> out of her purse and checked her lipstick.
- Ⓓ The <u>compact</u> arrangement of her things in the corner suggested she had not brought much with her.

---

13 She worked hard on her <u>own</u> paper and did not want anyone else to steal it.

- Ⓐ He did not <u>own</u> his home but was renting from a woman down the street.
- Ⓑ She could hold her <u>own</u> when lifting weights with the boys.
- Ⓒ He was glad to have his <u>own</u> money for a change.
- Ⓓ She was on her <u>own</u> in a foreign country and she was both scared and excited.

Directions: Read each set of two sentences. Then choose the one word that fits in both sentences.

Example:

The balloon rose because it was lighter than _____.

I don't know what time the TV show is going to _____.

- Ⓐ clouds
- Ⓑ air
- Ⓒ heaven
- Ⓓ close

Answer:

- Ⓑ air

---

14 She is going to _____ a notice on the bulletin board asking for information about the missing dog.

He installed a _____ to carry the telephone lines.

- Ⓐ post
- Ⓑ place
- Ⓒ wire
- Ⓓ string

---

15 She wore a _____ of pearls as her only adornment on her wedding day.

Annette used _____ to tie the packages into her trunk.

- Ⓐ rope
- Ⓑ twine
- Ⓒ string
- Ⓓ group

(See page 113 for answer key.)

Words in Context

Students understand word meanings in context when they can determine word meanings based on the information in the sentence or paragraph around them. The meaning can be gleaned from clues given in the passage if the reader pays careful attention to these clues.

In order to determine the meaning of words in context, the student will need to pay careful attention to the passage and notice key words and phrases. Knowing how to find these key components is the first step to finding the right answer. Students who are good at reading comprehension are usually good at this skill as well. Thus, many of the strategies discussed in Chapter 3 will also apply here.

Let's consider a passage with some new vocabulary words and plan a strategy for how your child can figure out word meanings from context clues. Look at this paragraph:

When a hurricane is about to strike the coast, it's often necessary for state governors to order *mandatory evacuations* of affected areas. Residents living in specified locations are ordered to leave their homes for safer destinations. This massive *exodus* of residents can create huge problems for law enforcement and highway department officials.

We'll focus on the meanings of the words *mandatory, evacuations,* and *exodus.* Try this step-by-step plan for discovering these word meanings:

1. Read the paragraph and then choose a few key words. Look especially at simple subjects and predicates. In our sample paragraph, these words include: *hurricane, strike, residents, ordered, create, safer, leave,* and *homes.* By reading this list, you can guess that the paragraph is about increasing residents' safety when a hurricane approaches.

2. Look at the sentence or phrase just after the word in question to see if there is a hint to word meaning. In this paragraph, the sentence after *mandatory evacuation* actually defines those words by saying, "Residents … are ordered to leave …"

3. Now read the paragraph and omit the target words. See if you can substitute other words and phrases that make sense. That is, use your guesses about the definitions in place of the target words.

4. If these words or phrases work, then you have most likely defined your target words.

What Tests May Ask

On a standardized test that assesses words in context, a student will be asked to read a sentence or paragraph that contains blanks and several choices of words to fit in the blanks. The student chooses the best word to fit in the blank.

What You and Your Child Can Do

There are several good ways you can work on words in context with your sixth grader.

Read. Read out loud to your child, and stop when you come to an unfamiliar word. Ask for guesses on what it means. Discuss clues in the passage that help your child find the meaning; then reread the passage and insert the meaning your child selected. Ask her if it makes sense based on the remainder of the passage.

Play Radio Tag. While riding in the car, listen to the radio and call attention to any unfamiliar words. Ask your child if she has heard them before. Try to find clues in the show and guess at the meaning of the word. Write the word down and check the meaning of it in a dictionary later.

Be a Model. Sixth graders often assume adults know just about every word they hear. Show your child that this isn't the case and that you're always learning new words. When you hear an unfamiliar word in conversation, ask about its meaning in front of your child. If you run into a new word while you're away, bring the word home and discuss the meaning in front of your child. Look it up in the dictionary together.

Keep a List. Keep a list of new words that you've learned from context. Challenge each family member to bring home one new word every week.

Pay Attention. Be aware of new words in places where we often ignore them. For example, menus often have words indicating cooking style, toppings, seasonings, or special ingredients.

Practice Skill: Words in Context

Directions: Read the following paragraph from *Tales from Shakespeare* by Charles and Mary Lamb. Choose the correct words to fill in the blanks from the selections below.

The two chief families in Verona were the rich Capulets and the Montagues. There had been an old ____16____ between these families, which was grown to such a height, and so deadly was the enmity between them, that it extended to the remotest ____17____, to the followers and retainers of both sides, insomuch that a servant of the house of Montague could not meet a servant of the house of Capulet, nor a Capulet ____18____ with a Montague by chance, but fierce words and sometimes bloodshed ensued; and frequent were the ____19____ from such accidental meetings, which disturbed the happy quiet of Verona's streets.

16　Ⓐ　discussion
　　Ⓑ　quarrel
　　Ⓒ　agreement
　　Ⓓ　house

17　Ⓐ　hills
　　Ⓑ　world
　　Ⓒ　kindred
　　Ⓓ　teachers

18　Ⓐ　encounter
　　Ⓑ　fight
　　Ⓒ　dance
　　Ⓓ　hike

19　Ⓐ　lights
　　Ⓑ　separations
　　Ⓒ　brawls
　　Ⓓ　agreements

(See page 113 for answer key.)

Word Study

Word study skills are used when students analyze words for their origins and meanings by looking at word parts. Attention is given to word endings and roots that give information about word meanings.

What You and Your Child Can Do

Parents who want to help their children with word study skills may find it helpful to:

• Point out words that originate from the same root words, such as:

aquatic, aquamarine, aquarium

biology, biosphere, biography

linguistics, sublingual, bilingual

socialization, socialist, society

• Involve your child in the study of a second language. Note words that are similar in English to those in the second language.

Practice Skill: Word Study

Directions: Choose the best answer.

Example:

Which of these words comes from the Latin word <u>laxare</u>, meaning to slacken?

　　Ⓐ　rind
　　Ⓑ　relax
　　Ⓒ　run
　　Ⓓ　rodent

Answer:

　　Ⓑ　relax

20 Which of these words comes from the Latin word <u>relevare</u>, meaning to lift, raise up again, or lighten?

(A) rent
(B) remainder
(C) relieve
(D) relic

21 Choose the answer that best defines the underlined part of the two words.

Normal<u>ize</u>   Trivial<u>ize</u>

(A) to minimize
(B) to make something _____ (in this case, either normal or trivial)
(C) to enlarge
(D) a person who _____ (in this case, either makes normal or trivial)

(See page 113 for answer key.)

Word Definitions

A student learns definitions by some of the same strategies already discussed. In most standardized tests assessing word definitions, students are given several definitions and must choose the one that matches the target word.

What You and Your Child Can Do

Help students learn word definitions by using some of these strategies:

• Make the dictionary a very important book in your house. Make sure you have a student dictionary, an unabridged dictionary, and a small paperback dictionary you can take along in the car on trips.

• Discuss word meanings with your child. Whenever a new word comes up, develop a definition and discuss ways the word is used.

• Play the Definition Game: Divide your family into two teams; you (as moderator) choose a very unusual word from the dictionary. One team must try to come up with a definition, and the other team must decide whether that definition is correct.

• Students often have school assignments to look up vocabulary words in dictionaries or glossaries. Know the words your child is studying and use them in conversation during the week.

Practice Skill: Word Definitions

Directions: Read each item. Choose the answer that expresses the definition of the underlined word.

Example:

a <u>thrifty</u> shopper

(A) quick
(B) frugal
(C) extravagant
(D) hurried

Answer:

(B) frugal

22 records from <u>antiquity</u>

(A) an ant farm
(B) an antique shop
(C) long ago
(D) relatives, such as aunts

23 a <u>perplexed</u> expression

(A) serene
(B) happy
(C) comfortable
(D) confused

(See page 113 for answer key.)

CHAPTER 3

Reading Comprehension

Tests of reading comprehension measure how much a student understands and remembers what he reads. Students are required to read critically, to make inferences about what they read, and to understand the literal information presented in the text.

What Your Sixth Grader Should Be Learning

By the sixth grade, your child should have a solid ability to read and to understand what he has read. He will be expected not only to read material, but to make inferences from what he reads, to know the difference between fact and opinion, to be able to compare and contrast. He should understand different writing genres and literary devices and be able to choose a title that applies to what he has read.

What Tests May Ask

Tests of reading comprehension are usually presented in multiple-choice format. Students must read a passage and then answer questions about what they've read. More and more, these tests also include questions that require students to write responses in either short answer format or complete sentences.

Reading comprehension items for sixth graders measure a range of skills, including:

- making comparisons
- knowing the difference between fact and opinion

- understanding sequence
- identifying details and drawing conclusions
- using a story web
- deriving word meaning
- identifying and understanding characters
- choosing reference sources and a good title
- generalizing, predicting, and inferring
- understanding the author's purpose and main idea
- interpreting persuasive language
- understanding literary devices and genres
- understanding cause and effect
- interpreting figurative language

What You and Your Child Can Do

Making sure a child comprehends what he reads is probably the most important job schools and parents do, but it's not very hard. Reading comprehension strategies are the most obvious and in some ways the easiest to implement. In order to increase reading comprehension, you'll need to:

1. Encourage your child's enjoyment of reading and speaking.
2. Encourage your child to read daily.
3. Increase the fluency with which your child reads.
4. Increase the vocabulary your child understands or is able to figure out.

5. Increase the pool of information with which your child is familiar.

6. Increase your child's ability to think and reason with words.

Step One: Encourage Your Child's Enjoyment of Reading and Speaking. It's no secret that we tend to do things more often when we enjoy them. Reading may seem like drudgery when students are first learning, and they may develop negative feelings about it. It's especially rough when students see reading as only useful in school, for subjects in which they have no inherent interest. Reading actually happens throughout the day, and it's essential for everything from writing a business plan to bird-watching. Helping children experience payoffs early on will increase reading comprehension quickly.

Start by thinking of your child's interests. Focus your selection of books in these areas, even if the books are easier than those your child reads in school. If your child is interested in sports, choose books related to sports. If your child enjoys humor, choose the funny papers or comic books. Talk about these books and the information they contain.

Don't forget nonfiction! Remember that in standardized tests, students will be tested on nonfiction writing as well as on passages of fiction. Make sure your child reads some expository writing (writing that contains facts, observations, and explanations, much like a science or social studies textbook).

Sixth graders may seem frighteningly mature at times, but don't forget that they still enjoy being read to. Many sixth graders can think on an adult level, but they may not be able to read on an adult level. If you read to them, they'll enjoy and understand more difficult information and may find it challenging and interesting. As you read, remember these pointers:

- Sixth graders avoid things they perceive as babyish. Choose topics that are challenging and complicated.

- Set aside some time each day for oral reading. Your child may enjoy reading orally for a while and having you read the remainder.

- Discuss what you read. The important ideas in a particular passage may not be obvious to a sixth grader. It's surprising what students don't understand because they miss the meaning of one word.

When you do discuss what you've read aloud, don't miss the opportunity to share your own thoughtful statements that can lead to further discussion. Consider the following dialogue:

DAD: This paragraph really touched me because I remember my dad talking about World War II. It was a very sad time for a lot of people.

SAM: Why was that?

DAD: Well, the Germans invaded other countries and especially persecuted Jewish people. They arrested Jews and sent them to concentration camps, just the way it happened in this story.

SAM: What's a concentration camp?

And the discussion continues. When Sam encounters a reference to concentration camps in the future, he'll likely respond with much more insight than before. Even though his teacher may have talked about this information many times in the past, learning doesn't always stay with the student unless it's related to things he knows and it's repeated at key times. Parent–child reading and discussing accomplishes just that.

Step Two: Encourage Your Child to Read Daily. Many teachers and schools have adopted daily reading requirements. Students are required to read a certain number of minutes each day, to discuss what they have read with a parent, to record what they have read, and to turn the record in on a weekly basis.

If your child's school hasn't implemented such a requirement, you may want to ask them to consider it; it will be much easier for you to insist on daily reading if the school requires it.

If your school doesn't demand daily reading, you can still require it at home. Remember the following in helping your child choose what to read:

- Reading must be fun in order to be enjoyed. Choose materials your child will like.

- Reading for pleasure may include material that is easier than what your child normally reads in school; that's okay.

- Reading out loud is good for students, but may be frustrating if the student has to read an entire passage.

- Listening to books read out loud helps students, because they listen to the flow of language and develop skills at assimilating facts and thinking about ideas.

It's very important that sixth graders work on complex reasoning skills. You can begin to do this by selecting some challenging reading material, but you can also discuss challenging ideas from simple texts. Likewise, you can use adult materials to illustrate some of the ideas you wish to communicate.

You may want to consider some books from reading lists developed especially for sixth graders. Ask your child's teacher for suggestions or visit your local library. *The way you approach reading is more important than the specific books you choose.*

TIP

Plan trips that include or focus on reading, such as to the library, the zoo, the museum, or historical sites. Read the information in exhibits with your child and make comments that bring the information home to your child.

Step Three: Increase the Fluency with Which Your Child Reads. Reading fluency is the rate and smoothness with which your child reads. This may not seem to be much of an issue, but it becomes very important in standardized testing

if your child reads much more slowly than others and doesn't finish the test. It's also true that the students who read fluently maintain the meaning of the passage and remember what they have read more frequently than do those who stammer and halt.

Some readers struggle with reading fluency for years and others seem to read fluently from the very beginning. If your sixth grader is among the lucky ones, be thankful and move on to other suggestions. If your child is less fluent, you may pick up some handy hints here:

1. Provide your child with fluent readers as role models. Read out loud to your child.

2. Listen to your child read. Does he miss the easy words or stumble on words he has seen many times before? Often it's the simple words that cause the problems. As you listen to your child read, note which words he seems to stumble on and make a list of them.

3. Break this list down into shorter lists and practice reading them out of context with your child. Do this for short periods each day. Include no more than 20 words each day.

4. Have your child practice with a passage that's easier than typical reading he does in school. Time him and note how many words he mispronounces.

5. Have your child read a familiar passage at least 3 times a week. Note each time how many words your child is able to read in a one-minute period.

6. Keep a graph of how this rate increases as time goes on. Your child may be interested in how his rate increases as his practice increases.

Step Four: Increase the Vocabulary Your Child Understands or Is Able to Figure Out. Reading is a task of understanding words in context. Consider the following sentences and see if you can understand them:

Tommy carried a *dibble* to school with him because he had to present an oral report on his gardening project.

Susan asked for a change of *venue* when she realized that the case had received so much attention.

Unless you understand the italicized words, your understanding of the sentences is inaccurate or incomplete. The same is true with students, who may misread an entire paragraph because of one word.

Most experts agree that the best way to increase vocabulary is to read lots of different types of writing. The first time you read a new vocabulary word it may well slip right by as you read. Almost no one learns new words from just one exposure; you need to see the word several times before you master it. *Especially in standardized testing, your child's lack of knowledge of one word may reduce or eliminate his comprehension of an entire passage.* Thus, every word he learns might increase his performance on a test.

The best ways you can help boost your child's vocabulary other than providing lots of reading opportunities are:

- Have your child participate in a wide variety of activities, including dance, gymnastics, music, fishing, or camping.
- Travel.
- Sample food and customs from other countries.

With each of these opportunities comes the chance to learn new words while also having fun. Every hobby or interest comes with its own special vocabulary. We can expose our sixth graders to new levels of vocabulary simply by taking them on excursions.

DAD: I'm surprised that no trout were biting my nymphs today.

ELLEN: What's a nymph?

DAD: It's a stage of insect development. I was using nymphs on my fly rod to try to catch trout today. Let me show you one.

ELLEN: I thought you were catching rainbows.

DAD: Yes, I was trying to catch rainbows. A rainbow is a kind of trout. Other kinds are brown trout, brook trout, and cutthroats.

This father could turn the discussion into a regular biology lesson if his child would allow it, or he could slip in little lessons when his daughter asked further questions.

Step Five: Increase the Pool of Information with Which Your Child Is Familiar. We understand a lot more of what we read when we understand something about the content of the story. For example, if you've ever been to a waterfall, you may understand why a character in a story might feel awe, fear, and amazement at the sight.

A person's fund of information is a result of his experiences, both those he's had in person and those he's experienced in books. Pay particular attention to the suggestions given here, because they all increase both vocabulary and fund of experiences. Encourage your child to:

Read the newspaper.

Visit historic sites.

Join Scouts and other organizations.

Study a second language.

Complete research projects.

Participate in summer camps and enrichment experiences.

Take music lessons.

Watch educational television shows.

Play with historic dolls or paper dolls.

Play games that emphasize facts and information.

Step Six: Increase Your Child's Ability to Think and Reason with Words. This skill is extremely important for sixth graders because it will be needed to answer questions requiring complex skills such as inferential reasoning and critical analysis.

Students use inferential reasoning when they must make conclusions about information based on facts they've been given. Critical analysis is required when students make judgments about information, people, or setting.

The ability to reason can be taught, but it's a very difficult area for many of us to tackle. Let's break it down into a few different areas of focus:

1. Determining the main idea of passages.

2. Noting patterns and trends (sequencing events, predicting outcomes, extending meaning, and making comparisons).

3. Analyzing information critically (differentiating fact and opinion, understanding feelings, drawing conclusions, applying story information, and understanding the author's purpose).

4. Identifying details.

1. Determine the main idea of passages. Finding the main idea is one of the first skills your child was taught in reading, and it's very important in standardized testing. To determine the main idea of a passage, your child should:

1. Survey the material. Look at the pictures. Read the captions under the pictures.

2. Read the main headings of the passage.

3. Read the first paragraph. Ask what the author is trying to communicate through this paragraph. There is usually a clue to what the main idea is either in this paragraph or in the final paragraph.

4. Read the remainder of the passage through quickly.

5. Read the last paragraph. Is the main idea restated, or possibly stated for the first time?

6. Reread the first paragraph. Ask yourself if your first guess about the main idea makes sense now.

As you and your child read together and try to find the main ideas in paragraphs, include questions from this list to discuss the material:

*   What did the author intend to tell us here?

*   How would you sum up the passage?

*   What would the characters say about what happened?

2. Note patterns and trends. Authors use a variety of techniques to call a reader's attention to a pattern in the text. The parts of the story help a reader analyze the patterns of importance; these parts include characterization, setting, plot, and use of symbols or metaphors.

Character may be used to convey repeated patterns by the author's descriptions of the characters, the actions the characters take, the feelings characters have about happenings in the story, or the changes the characters go through.

Setting is important in higher-level thinking because of the impact of the setting on the plot and character. A difficult, harsh environment causes characters to act in certain ways and gives them predictable conflicts to encounter.

Plot conveys an author's intent by the use of conflicts the characters encounter. The actions taken and reactions received from others frequently give the reader clues from which to make inferences. The sequence of events helps the reader make predictions about what happened before the story and what will happen after.

Recognizing the use of symbols and metaphors by the author is an important part of understanding the implications and theme of the story. A metaphor is an idea in the story that is used to represent or stand for another, more abstract idea. Sixth graders may begin to encounter metaphors such as:

a trip as a metaphor for a life;

a circle as a metaphor for the cycle of life, including the death and birth of beings; and

a year as a metaphor for the natural progression of living and dying.

Without becoming a literary expert, how can you help a sixth grader with these complicated ideas? Your training and expertise may be in other areas, and not in the analysis of thoughts and ideas in literature. Never fear! You can do it, but it does require some time and some thought. When trying to help your child note patterns and trends, consider asking questions like these:

- What probably happened before this passage?
- What will probably happen next?
- How were the people feeling at the time?
- What meaning does this story have for our lives?
- What would you have done in this situation?

3. Analyze information critically. Students analyze information critically when they make judgments about what an author says, draw conclusions from the facts, and determine whether the information is fact or opinion. It's often much easier for your child to develop these skills in a discussion group rather than by himself, so parents can play a big role in this area.

Parents have only a few times when they can work with their children on these skills, and we have to make the most of our opportunities. Sitting around the dinner table, driving in the car, and helping with homework are three times that come to mind. Discussions about information from other sources can also be used to develop critical thinking skills. These include:

- The news you hear on the radio
- Articles in the newspaper
- Adult conversations children overhear

Ask these questions to get your student to demonstrate skills in this area:

- What do you think about what happened?
- What applications does this information have for you?
- What does this mean in connection with what you already know?

- Does the author draw conclusions that are not warranted by the information?
- What conclusions would you draw from the same data?
- What questions would you ask about the data presented?

When working on reading comprehension, consider these ideas as well:

- It is much easier to help someone with reading comprehension if you have read the book yourself. Sixth-grade books are fun even for adults to read, and will be more fun for your child if he can discuss them with a parent. As you and your child read, you can explore interpretations together.
- Reading clubs composed of parents and children provide opportunities for discussion. In a group, you'll hear a variety of interpretations.

4. Identify details. Students often miss important details of a passage by reading the passage too quickly or by not paying careful attention to some of the words. On a standardized test, sixth graders may be asked about information contained in only one word or phrase.

You can help your child find the details of a story by helping him identify key words in a sentence. Remember that the subject and predicate of a sentence contain important keys to the thought, but the rest of the sentence may give the details that are important in questions.

Some students need to use a marker as they read to help them pay attention to each word or phrase. Others need to reread or scan passages. Remind your child that when questions are asked about passage details, it's okay to look back at the passage!

Practice Skill: Reading Comprehension

Directions: Read the following selection from *The Wind in the Willows* by Kenneth Grahame and answer the questions that follow.

"You knew it must come to this, sooner or later, Toad," the Badger explained severely. "You've disregarded all the warnings we've given you, you've gone on squandering the money your father left you, and you're getting us animals a bad name in the district by your furious driving and your smashes and your rows with the police. Independence is all very well, but we animals never allow our friends to make fools of themselves beyond a certain limit; and that limit you've reached. Now, you're a good fellow in many respects, and I don't want to be too hard on you. I'll make one more effort to bring you to reason. You will come with me into the smoking-room, and there you will hear some facts about yourself; and we'll see whether you come out of that room the same Toad that you went in."

He took Toad firmly by the arm, led him into the smoking-room, and closed the door behind them.

"That's no good!" said the Rat contemptuously. "*Talking* to Toad'll never cure him. He'll *say* anything."

They made themselves comfortable in arm-chairs and waited patiently. Through the closed door they could just hear the long continuous drone of the Badger's voice, rising and falling in waves of <u>oratory</u>; and presently they noticed that the sermon began to be punctuated at intervals by long-drawn sobs, evidently proceeding from the bosom of Toad, who was a soft-hearted and affectionate fellow, very easily converted—for the time being—to any point of view.

After some three-quarters of an hour the door opened, and the Badger reappeared, solemnly leading by the paw a very limp and dejected Toad. His skin hung baggily about him, his legs wobbled, and his cheeks were furrowed by the tears so plentifully called forth by the Badger's moving discourse.

1 In this passage, which of the following does the author suggest happened just before this event?

- (A) Toad had intentionally burned down the barn.
- (B) Toad had run away from home.
- (C) Toad had wasted money and driven recklessly in his car.
- (D) Toad had stolen a car.

2 Which role is the most similar to that taken by Badger in this passage?

- (A) supportive uncle
- (B) lecturing father
- (C) accepting teacher
- (D) teaching minister

3 In the story, the word <u>oratory</u> means:

- (A) a song
- (B) a speech
- (C) a long silence
- (D) sounds of water

4 At the end of the passage, the other animals notice that Toad appears to have been:

- (A) crying
- (B) laughing
- (C) thinking
- (D) talking

5 The best title for this selection is:

- (A) Badger becomes a Minister
- (B) Toad and Badger Are Friends
- (C) Toad's Automobile Accident
- (D) Toad's Lesson

6 Before Badger and Toad went into the smoking-room, the mood in the room was _____. After Badger and Toad came out of the smoking-room, Toad appeared to be feeling _____.

- (A) angry, happy
- (B) tense, remorseful
- (C) hopeful, angry
- (D) tense, contented

Directions: Read the following passage from *Treasure Island* by Robert Louis Stevenson and answer questions 7 through 11.

From the side of the hill which was here steep and stony, a spout of gravel was dislodged, and fell rattling and bounding through the trees. My eyes turned instinctively in that direction, and I saw a figure leap with great rapidity behind the trunk of a pine. What it was, whether bear or man or monkey, I could in nowise tell. It seemed dark and shaggy; more I knew not. But the terror of this new apparition brought me to a stand.

I was now, it seemed, cut off upon both sides; behind me the murderers, before me this lurking nondescript. And immediately I began to prefer the dangers that I knew to those I knew not. Silver himself appeared less terrible in contrast with this creature of the woods, and I turned on my heel, and, looking sharply behind me over my shoulder, began to retrace my steps in the direction of the boats.

Instantly the figure reappeared, and, making a wide circuit, began to head me off. I was tired, at any rate; but had I been as fresh as when I rose, I could see it was in vain for me to contend in speed with such an adversary. From trunk to trunk the creature flitted like a deer, running manlike on two legs, but unlike any man that I had ever seen, stooping almost double as it ran. Yet a

man it was, I could no longer be in doubt about that.

I began to recall what I had heard of cannibals, I was within an ace of calling for help. But the mere fact that he was a man, however wild, had somewhat reassured me, and my fear of Silver began to revive in proportion. I stood still, therefore, and cast about for some method of escape; and as I was so thinking, the recollection of my pistol flashed into my mind. As soon as I remembered I was not defenseless, courage glowed again in my heart; and I set my face resolutely for this man of the island, and walked briskly toward him.

He was concealed by this time behind another tree trunk; but he must have been watching me closely, for as soon as I began to move in his direction he reappeared and took a step to meet me. Then he hesitated, drew back, came forward again, and at last, to my wonder and confusion, threw himself on his knees and held out his clasped hands in supplication.

At that I once more stopped.

"Who are you?" I asked.

"Ben Gunn," he answered, and his voice sounded hoarse and awkward, like a rusty lock. "I'm poor Ben Gunn, I am; and I haven't spoke with a Christian these three years."

7 What do you think the narrator was doing at the beginning of this passage?

- (A) sitting on a hillside
- (B) running away from other people
- (C) walking in the hill country
- (D) rowing a boat

8 Silver is probably:

- (A) a companion of the narrator
- (B) a dark and shaggy creature
- (C) a horse
- (D) a murderer

9   The narrator's attention is first drawn to the creature by:

  (A) the call of a wild beast

  (B) the sight of a figure moving among the trees

  (C) the sound of rocks rolling down the hill

  (D) a vague sense of danger

10 Seeing the shaggy figure, the narrator feels:

  (A) fear
  (B) curiosity
  (C) indecision
  (D) both A and C

11 Why was Ben Gunn hiding behind the trees?

  (A) He was as afraid of the narrator as the narrator was of him.

  (B) He was hungry.

  (C) He was embarrassed by his appearance.

  (D) He was afraid of the murderers.

Directions: Read this passage of nonfiction from *Lives of the Hunted* by Ernest Thompson Seton. Use the information presented to answer questions 12 through 16.

Johnny was a queer little bear cub that lived with Grumpy, his mother, in the Yellowstone Park. They were among the many bears that found a desirable home in the country about the Fountain Hotel.

The steward of the hotel had ordered the kitchen garbage to be dumped in an open glade of the surrounding forest, thus providing, throughout the season, a daily feast for the bears, and their numbers have increased each year since the law of the land has made the park a haven of refuge where no wild thing may be harmed. They have accepted man's peace-offering, and many of them have become so well known to the Hotel men that they have received names suggested by their looks or ways. Slim Jim was a very long-legged thin Blackbear; Snuffy was a Blackbear that looked as though he had been singed; Fatty was a very fat, lazy Bear that always lay down to eat; the Twins were two half-grown, ragged specimens that always came and went together. But Grumpy and Little Johnny were the best known of them all.

Grumpy was the biggest and fiercest of the Blackbears, and Johnny, apparently her only son, was a peculiarly tiresome little cub, for he seemed never to cease either grumbling or whining. This probably meant that he was sick, for a healthy little bear does not grumble all the time, any more than a healthy child. And indeed Johnny looked sick; he was the most miserable specimen in the park. His whole appearance suggested dyspepsia; and this I quite understood when I saw the awful mixtures he would eat at that garbage-heap. Anything at all that he fancied he would try. And his mother allowed him to do as he pleased; so, after all, it was chiefly her fault, for she should not have permitted such things.

Johnny had only three good legs, his coat was faded and mangy, his limbs were thin, and his ears and paunch were disproportionately large. Yet his mother thought the world of him. She was evidently convinced that he was a little beauty and the Prince of all Bears, so, of course, she quite spoiled him. She was always ready to get into trouble on his account, and he was always delighted to lead her there. Although such a wretched little failure, Johnny was far from being a fool, for he usually knew just what he wanted and how to get it, if teasing his mother could carry the point.

12 The author of this passage would probably agree with which of the following statements?

Ⓐ It was a good idea for the hotel staff to leave out food scraps for the bears of Yellowstone, because during some years there was not enough food available in the wild.

Ⓑ The bears of Yellowstone were damaged by the practice of leaving food out, because some chose to eat food that was not healthy for them.

Ⓒ Yellowstone Park should probably not have been built because it interfered with the natural habitat of the bears.

Ⓓ The bears of the park were really a nuisance and prevented the park guests from enjoying the beautiful scenery without fear.

13 The author suggests that the bears in this passage have:

Ⓐ remained fearful of humans

Ⓑ become accustomed to humans

Ⓒ been aggressive and dangerous toward park visitors

Ⓓ become playmates of the humans

14 In the third paragraph, the author uses the word <u>dyspepsia</u>. What meaning is most likely for this word, given the other information presented in the paragraph?

Ⓐ broken leg
Ⓑ depression
Ⓒ indigestion
Ⓓ tooth decay

15 Which of the following is the best statement about the author's technique in the final paragraph of this passage?

Ⓐ The author applies human logic and experience to the observations he has made of the bear mother and her son, and makes guesses about the mother's thoughts.

Ⓑ The author makes accurate statements about the feelings of the bear for her son.

Ⓒ The author is attempting to be funny in his description of the bear and her cub.

Ⓓ The author is going off on a tangent that has no meaning for the bears or the reader of this passage.

16 Why did the hotel men name two bears "the Twins"?

Ⓐ They looked alike.
Ⓑ They acted alike.
Ⓒ They ate the same food.
Ⓓ They were always seen together.

(See page 113 for answer key.)

CHAPTER 4

Language Mechanics

"Language mechanics" sounds like a complex and somewhat daunting prospect, but in fact it simply refers to capitalization and punctuation. These are not subjects that your average sixth grader will find riveting, but with today's trend toward more and more standardized tests, you can expect plenty of emphasis on the basic mechanics of English both in the classroom and in the testing situation.

What Your Sixth Grader Should Be Learning

Your sixth grader will be expected to know how to use all forms of capital letters and punctuation, including the correct use of punctuation endings, commas, quotation marks, and apostrophes. Forming the possessive can often be tricky, but sixth graders should have mastered the basic rules.

However, don't be surprised if your sixth grader has trouble with these skills. (Lots of adults still aren't sure of many of the mechanics of English either.) Many creative students struggle with the specifics; this shouldn't surprise you, although it may be frustrating to you both. It's important to remember that writing is a very complicated process and that students may become frustrated with it if they get lots of negative feedback.

What Tests May Ask

Because language mechanics is becoming an important part of the curriculum in many schools, standardized tests of these skills are designed to measure a student's understanding of the rules of punctuation and capitalization. In some cases, students will be asked to edit incorrect sentences and paragraphs; in other cases, they'll be asked to choose the correct form of a sentence, phrase, or paragraph. More and more tests include items requiring students to produce sentences and paragraphs, so your child should be ready to generate her own properly capitalized and punctuated passages.

In this chapter, we focus on the specifics needed to write a passage that meets standard rules and to recognize the work of others that has used the rules correctly. These skills are best taught in the context of a written passage, and the strategies for increasing written productivity discussed in Chapter 5 will be helpful here too.

What You and Your Child Can Do

First of all, it will help if you have a general idea of the mistakes your child is making when it comes to capitalization and punctuation. Look over your child's homework and analyze any errors you find. Let's say you notice she can capitalize fairly well and uses end punctuation reliably, but has problems in forming the possessive and placing end punctuation when quotes are used.

These are the areas, then, you'll need to zero in on. Eventually, you want to have your child use apostrophes and quotation marks so automatically she doesn't even have to think about it.

Here are a few tips on how to provide positive feedback and avoid frustrating your child:

- Don't try to teach all the punctuation and capitalization strategies at once.

- Focus on a few examples at a time.

- Many adults aren't sure about proper capitalization and punctuation. When this happens to you, don't fake it! Admit you're not sure and consult a reference book or call the teacher.

- Encourage your child to develop a first and second draft before writing the final report or project. This provides more opportunities for gentle feedback and reinforcement.

If apostrophes and quote marks give your child a problem, hand her the evening newspaper and a couple of markers and have her circle all the apostrophes and quote marks she can find on the front page.

If you have a computer, type her a letter and mix up the punctuation. Have her go through and correct what you've done. For a change, allow her to use your computer grammar checker. See if she can spot the problem before the computer does. Many children who balk at doing paper-and-pencil punctuation drill have no problem participating in electronic practice.

Let your child try on the "editor hat": Write out a number of sentences missing important capitalization or punctuation. Set a timer and have your child go over your "news article." Give your child a small reward (piece of candy, fruit, or a penny) for every error she catches.

Another good way to work on punctuation and capitalization is to play games. Most parents and their children seem to spend lots of time driving around in the family car, especially when taking family vacations. You can make the most of this time by playing games that focus on academic skills.

Read road signs to see if they use the rules of capitalization and punctuation correctly. Give points to each person who can find and correct an error. Watch for license plates from different states and provinces. Have the student make a list of states, with appropriate capitalization. Challenge your child to name the state capitals to earn extra points. Have your child write the names of the states and capitals appropriately.

Capitalization

You may think that your sixth grader has been working on capitalization since kindergarten, but it's not always as easy for kids as it may seem. Sixth graders still may make mistakes in this area; but because all standardized tests assess capitalization, you'll need to make sure your child has a thorough understanding of the rules.

What Your Sixth Grader Should Be Learning

Your sixth grader should be thoroughly comfortable with the use of capital letters, not just for the beginnings of sentences but for book and movie titles, the personal pronoun *I,* and all proper nouns.

What You and Your Child Can Do

Make sure that your child really understands what and when to capitalize. Glance over schoolwork and homework to make sure all the rules are being followed.

If you notice a problem, have your child practice a few sentences each night containing words that she has been having problems with.

Punctuation

If you've ever gotten a business letter replete with punctuation mistakes, you know how damaging a poorly written letter can be. The earlier your child learns and understands correct punctuation, the more naturally she'll be able to incorporate good punctuation in her work as she gets older. Punctuation rules aren't very hard; with a bit of practice, your sixth grader should be able to punctuate like a pro!

What Your Sixth Grader Should Be Learning

In sixth grade, students are expected to understand and be able to use proper punctuation marks in sentences, including periods, question marks, exclamation points, quotation marks, and apostrophes. By paying more attention to these important little marks, they'll improve their performance in daily writing and in standardized tests.

What You and Your Child Can Do

Don't be upset if you can't remember all the punctuation rules. Many adults are rusty in this department—but it's not hard to brush up on the basics. The rules of punctuation haven't really changed over the years, so if you can find an old grammar book (or borrow one from the library) you'll be able to help your child brush up on her skills. There are several strategies you can try with your sixth grader to increase awareness of all types of punctuation:

- Use a highlighter to mark end punctuation marks in a variety of sentences. Choose a family letter and have your child read the letter out loud and indicate what kind of sentences are in the letter and what the end punctuation should be.

- Write a letter with no end punctuation marks in sentences. Have your student go through the letter and mark in the punctuation marks with a colored marker.

- Choose a passage from a favorite book. During a family trip, read this passage out loud and stop at the end of sentences. Have your child supply the appropriate punctuation marks.

- Students often have problems with commas and may tend to follow the old adage, "When in doubt, use a comma." In fact, good writers have learned just the opposite: "When in doubt, take it out." When your child has removed as many commas as possible, have her read the sentence out loud. Spoken language reflects the pause needed for a comma; reading a sentence out loud will reveal where the commas should go.

- After students have written a paragraph, reread it with them and mark punctuation marks with a colored pencil or marker. Even marking correct punctuation is a good idea because it will call the student's attention to the mark.

- Have your child write a formal (or business) letter to an organization or person of interest, such as a fan letter to a favorite author or sports star or a request for tickets to a baseball game or for brochures about a vacation destination.

Practice Skill: Punctuation and Capitalization

Directions: Choose the punctuation mark that is needed in each of the following sentences. Choose "None" if no more punctuation marks are needed.

Example:

> Where are you going on vacation this summer
>
> (A) !
> (B) .
> (C) ?
> (D) None

Answer:

> (C) Where are you going on vacation this summer?

1   Our school was built several years ago
when the old school building became too
small for the students

(A)   !
(B)   .
(C)   ?
(D)   None

2   Why did they choose to build the building so far out in the country

(A)   !
(B)   .
(C)   ?
(D)   None

3   Along with other members of the community, we decided to build on that
property

(A)   !
(B)   .
(C)   ?
(D)   None

4   "Our community had planning sessions
to decide where to build the school"
said the principal.

(A)   ,
(B)   .
(C)   ?
(D)   None

5   I would really like to read books about
fish mammals and birds.

(A)   ,
(B)   .
(C)   ;
(D)   None

6   Why don't you go to the library?" her
teacher asked.

(A)   ,
(B)   .
(C)   "
(D)   None

Directions: Choose the answer that shows
correct punctuation and capitalization.

Example:

(A)   I think we are spending the summer in paris france.

(B)   "Do you know where we are", Seth?

(C)   Using great balance, the gymnast
performed an arabesque on the
beam.

(D)   "Look out!" he cried

Answer:

(C)   Using great balance, the gymnast
performed an arabesque on the
bars.

7   (A)   We live in a town that has beautiful
mountains valleys and lakes near-
by.

(B)   I wish we could travel to places we
have read about in the magazine
my family has

(C)   "Someone in our family just
returned from a long vacation in
the mountains," Annie stated.

(D)   Ellen and her friends went to
Athens Georgia for a gymnastics
meet.

8 Ⓐ We like to listen to some very interesting radio shows

   Ⓑ When you have an illness, you should drink lots of liquids visit your doctor, and follow instructions.

   Ⓒ Dad, are you coming home early today?

   Ⓓ I am going to see Dr Taylor for a physical examination today.

Directions: Choose the correct way to write the underlined part of the sentence.

Example:

"Let's get home <u>soon Susan</u> said.

Ⓐ soon", Susan
Ⓑ soon," Susan
Ⓒ soon, Susan
Ⓓ soon" Susan

Answer:

Ⓑ "Let's get home soon," Susan said.

---

9 We are out of school <u>today but</u> we still have to do our homework.

Ⓐ today. But
Ⓑ today, but
Ⓒ today but,
Ⓓ Correct as it is

---

10 "Let's go to the game <u>together suggested</u> Bette.

Ⓐ together. Suggested
Ⓑ together, suggested
Ⓒ together," suggested
Ⓓ Correct as it is

---

11 We need to go to Mr. <u>Maloneys</u> garden.

Ⓐ Maloney's
Ⓑ Maloneys'
Ⓒ Maloneys"
Ⓓ Correct as it is

Directions: Look at the underlined parts of the following paragraph and then answer questions 12–14, choosing the correctly capitalized and punctuated phrase.

One of the most <u>famous americans of the twentieth</u> century was Dr. Martin Luther King. Dr. King was born on January 15, 1929. He lived most of his life in Atlanta, Georgia, <u>where he was a minister.</u> He is most famous for his advocacy of civil rights for African Americans. His speech, "I Have a Dream," is familiar to most <u>americans and has been</u> an inspiration to people who share his desire for equal treatment of all people.

12 Ⓐ famous Americans' of the twentieth

    Ⓑ famous Americans of the twentieth

    Ⓒ Famous Americans of the Twentieth

    Ⓓ Famous Americans of the twentieth

---

13 Ⓐ Where he was a minister
    Ⓑ where he was a minister?
    Ⓒ where he was a minister.
    Ⓓ where he was a Minister.

---

14 Ⓐ Americans and has been
    Ⓑ americans, and has been
    Ⓒ Americans, and has been
    Ⓓ Americans' and has been

Directions: Read the following letter and answer questions 15 through 17.

101 Main Street
<u>lexington, s. c. 29072</u>
January 25, 2000

Camp Rocks for Boys
35 Hilly Way
Highlands, N. C. 28741

<u>dear mr. ramey:</u>

I am interested in obtaining information about your camp that focuses on camping and backcountry skills. I am 13 years old and have never done any camping or <u>back-packing but I</u> am in very good physical condition and am motivated to learn. Could I possibly participate?

I look forward to hearing from you soon. If you have questions, please reach me at the above address or by phone at (803) 555-9999. Thank you for your help.

Sincerely,

Trey Camper

15 Ⓐ Lexington SC 29072
   Ⓑ Lexington, S. C. 29072
   Ⓒ Lexington sc 29072
   Ⓓ Lexington-SC 29072

16 Ⓐ Dear Mr Ramey:
   Ⓑ Dear Mr Ramey,
   Ⓒ Dear Mr. Ramey
   Ⓓ Dear Mr. Ramey:

17 Ⓐ backpacking but I
   Ⓑ backpacking; but I
   Ⓒ backpacking, but I
   Ⓓ backpacking. But I

Directions: For questions 18 and 19, read the sentence with a blank. Choose the answer that completes the sentence and uses correct capitalization and punctuation.

Example:

I live in _____.

Ⓐ boston Massachusetts
Ⓑ Boston Massachusetts
Ⓒ Boston Massachusetts
Ⓓ Boston, Massachusetts.

Answer:

Ⓓ I live in Boston, Massachusetts.

18 _____ will be our guest at next week's Homecoming celebration.
   Ⓐ Rep. Woodley
   Ⓑ Rep Woodley
   Ⓒ Representative. Woodley
   Ⓓ Mr. Representative Woodley

19 We enjoy going to_____ with our family each summer.
   Ⓐ Cherry grove beach, South Carolina,
   Ⓑ Cherry Grove beach, South carolina
   Ⓒ Cherry Grove Beach, South Carolina,
   Ⓓ Cherry grove beach, South carolina,

Directions: Read this story and choose the correct capitalization for the underlined words in questions 20–23.

Playing in the surf is one of the best parts of a beach vacation for me. I especially enjoy the warm feeling of the waters of the <u>atlantic ocean.</u> If we go in <u>june or july</u>, there are lots of other people there enjoying the beach and the sunshine. Sometimes, I swim for over an hour <u>with aunt bette and uncle fred.</u> It is most fun, though, when my <u>little cousins</u> are there because they get really excited to see the waves.

20 Ⓐ Atlantic ocean
   Ⓑ atlantic ocean
   Ⓒ Atlantic Ocean
   Ⓓ atlantic Ocean

21 Ⓐ june or july
   Ⓑ June or July
   Ⓒ June or july
   Ⓓ june or July

22
  Ⓐ Aunt bette and Uncle fred
  Ⓑ Aunt Bette and uncle fred
  Ⓒ My Aunt Bette and My Uncle Fred
  Ⓓ Aunt Bette and Uncle Fred

23
  Ⓐ little cousins
  Ⓑ Little cousins
  Ⓒ Little Cousins
  Ⓓ little Cousins

Directions: Choose the correct way to finish these sentences.

Example:

She put dog food into her _____ dish.

  Ⓐ dogs
  Ⓑ dog's
  Ⓒ dogs'
  Ⓓ dog

Answer:

  Ⓑ She put dog food into her dog's dish.

24 I really enjoy going to my _____ house after school.

  Ⓐ friends
  Ⓑ friend's
  Ⓒ friend
  Ⓓ friends'

25 Ellen and Annie especially like to go to the beach. Their mother would probably say it is one of the _____ favorite destinations.

  Ⓐ girls
  Ⓑ girl's
  Ⓒ girls'
  Ⓓ girls's

(See page 113 for answer key.)

CHAPTER 5

Language Expression

Language expression includes what most people think of as English grammar, including items measuring use of parts of speech, sentence construction, paragraph form, and word usage.

What Your Sixth Grader Should Be Learning

By the sixth grade, your child should be completely comfortable with all forms of basic grammar, and be able to write fluently using accurate parts of speech, good paragraphs, and logical sentence construction. He should be able to identify all forms of grammar and be able to write using correct agreement between nouns and verbs. He should be able to identify a verb, noun, adjective, and adverb and to understand that words can be used as several different parts of speech.

What Tests May Ask

Standardized tests today are designed to measure a student's skills at using words to express ideas. Tests will ask your child to edit sentences for grammatical errors, combine simple sentences into complex ones, and make sure subject and verb tenses agree. They will ask your child to identify different parts of speech in examples; identify topic sentences; and use past, present, and future verb tenses.

Although the format of many questions in the past has been largely multiple-choice, tests increasingly include items requiring the student to write a sentence or paragraph. Therefore, your child should be prepared to complete both types of items.

Certainly, the more important skill for life and future education is that of actually composing and writing a sentence and passage, and this skill is emphasized in testing.

What You and Your Child Can Do

If you want your child to speak and write correct English, you'll need to model those behaviors yourself. When you hear your child make a grammatical mistake, gently correct him:

CHILD: Emma and me went to the mall to get new clothes for the dance.

YOU: Emma and I ... remember, if you take away the "Emma" in that sentence, it wouldn't make any sense to say "... me went to the mall."

CHILD (laughing): Okay! Emma and I went ....

Some children like to play editor and enjoy a timed contest: Everyone takes a different colored marker and sits down with a page from the newspaper. Pick a part of speech (noun, verb, adjective) and for 30 seconds everyone circles each instance of that part of speech. When time is up, the winner is the one who's found the most examples. This is particularly helpful practice because standardized tests primarily test your child's ability to spot and identify parts of speech and grammar mistakes, which is quite a different skill from coming up with your own examples.

Grammar Tools

Grammar tools are the building blocks of writing. The best ways to improve writing skills are to write and to read other people's writing. It's not surprising that the best readers are frequently good writers, because we learn about the way information is communicated by seeing it in print. Therefore, many of the reading skills are also applicable to grammar and writing.

How can you help your child develop these skills? There are seven primary ways of doing this.

1. Think of ways your child can use writing every day to accomplish things he wants to do.

2. Encourage the habit of writing every day.

3. Provide the materials and space your student needs to write.

4. Provide encouraging feedback about the writing your child produces.

5. Provide corrective feedback.

6. Let your child see you write every day.

7. Look at writing together and talk about the style.

Use Everyday Writing. Have your child write thank-you letters, invitations, ticket requests, fan letters, daily journals, a private diary, grocery lists, catalog orders, Christmas or birthday lists, and notes for teachers.

Many parents and employees now rely almost exclusively on computers to write letters and documents. Students have access to computers in schools, but frequently the access is some-

what limited and is not the primary method of writing in school. Standardized tests that have writing items usually do not have computers on site to allow students to type in their answers. There are certain exceptions, for instance when students have diagnosed disabilities, but the majority of students will be producing their work by hand for standardized testing at this point.

Provide Encouragement. Many students who aren't good at writing and grammar avoid it because they're afraid their writing may be unacceptable. This fear is often related to how adults have responded to their writing in the past, and it can be overcome by your careful comments. It may be especially hard for sixth graders to overcome this, because many have received negative feedback from parents and teachers for several years.

Positive feedback doesn't mean saying that everything is good or well done. Rather, it means primarily engaging in the process and taking time to understand and comment on the work.

- Always read every word of what your child has written. If writing has become difficult for your child, he needs to know that you see his writing as important and good.

- Ask your child to clarify what he has written.

- React to the passage appropriately. Laugh if the subject is funny. Indicate appreciation if he has written something for you. Nod as you read.

- Restate what the passage has said to you. Sometimes students don't know they've left something out or used an incorrect verb form that's critical to the comprehension of the passage.

- Save some examples of your child's work so you can show him how much progress he has made since beginning to write.

The process of encouraging writing is not very different from encouraging other activities. The primary differences are that you may be working against a history of failure and that you may encounter resistance to the process. Remember that the sensitivity of your comments will determine the future progress of your efforts.

Providing corrective feedback is hard for many parents. Teachers often have certain ways that they teach grammar and writing. Many use a series of steps to develop ideas and organize thoughts. Check with your child's teacher to find out if there is a writing plan. It might involve these four steps:

1. *Get ideas.* Idea generation is a very important step. The quality of a written product is partly dependent upon what is communicated, not just on whether the appropriate capitalization and punctuation is used. Help your child generate ideas by:

   - **Brainstorming**—Try brainstorming open-ended ideas and writing these down.

   - **Making a web**—The child puts a central idea in a middle circle and draws lines to connect other ideas to the central idea.

   - **Outlining**—An outline can be used to put the ideas in a logical sequence.

2. *Identify sequence.* After your child has generated some ideas, help your child identify the sequence in which the ideas should be presented. Make notes about this sequence.

3. *Write topic.* Have your child write a sentence that clearly states the topic.

4. *Add related ideas.* List several ideas connected to the topic. If the writing is to include only one paragraph, then you're finished. If there are multiple paragraphs, then begin with the second paragraph and follow the same procedure.

After your child has finished his first draft, ask him to reread it and correct any errors he sees. Read it out loud yourself and let him listen. The process of listening to his own work may help him identify other errors he didn't see

the first time through. Read the passage again silently and choose any areas you think need special focus. Give a few helpful hints on these.

TIP

If your child has significant problems with grammar and writing, you may have to select only one or two points to make about any piece. When you begin to work on writing, remember that the point of writing is to communicate information. Students must generate ideas before they can begin to communicate.

If your child can come up with ideas out loud but doesn't produce them in writing, use strategies such as note-taking, tape-recording, or transcribing (you write the ideas down as your child talks). It's a slow process that may be frustrating if not attempted in a calm, nondemanding way. Again, remember that writing takes a long time and that a product may not be finished on the same day it's begun.

What Tests May Ask

It's still true that most of the time, standardized tests give your students multiple-choice questions in order to assess knowledge of grammar. It's a somewhat different task to recognize information versus writing it, and good strategies will include practice in both.

Have your child try to find errors in things adults produce. This is fascinating to most sixth graders, but you have to be careful how the student handles the information. Help him realize that all people make mistakes but that the mistakes we make in writing are preserved for all to read and critique. Students should be encouraged to be tactful when finding errors, as they may find them in some very unlikely places!

Practice Skill: Grammar Tools

Directions: Read the directions for each section. Choose the answer you think is correct.

For questions 1–3, choose the word or phrase that best completes the sentence.

Example:

His dog was the _____ retriever at the dog show.

(A) prettier
(B) prettiest
(C) most pretty
(D) pretty

Answer:

(B) His dog was the prettiest retriever at the dog show.

1 We wonder ____ will run for Class President.

(A) whom
(B) who
(C) whether
(D) whoever

2 The work he completed in class was the _____ he had ever done.

(A) more accurate
(B) most accurate
(C) accuratest
(D) accurate

3 Katie _____ in her room the entire afternoon.

(A) were playing
(B) are playing
(C) was playing
(D) play

Directions: For questions 4–7, choose the answer in which a usage mistake is made.

Example:

(A) Him and I went
(B) down to the park

(C) to see the ducks.
(D) No mistakes.

Answer:

(A) Him and I went

4 (A) Last week, I seen some students
(B) from our high school
(C) at a football game.
(D) No mistakes.

5 (A) The students in the cafeteria didn't want
(B) none of the birthday cake that had been sitting around
(C) for several days.
(D) No mistakes.

6 (A) When we returned to her house after the game,
(B) we grab a coat
(C) and raced out to the restaurant.
(D) No mistakes.

7 (A) I noticed that the fifth graders
(B) were the quieter class in the
(C) entire school that day.
(D) No mistakes.

Directions: For questions 8–10, choose the sentence that is complete and correctly written.

Example:

(A) Running down the road.
(B) Eyes ablaze, she faced her tormentors.
(C) Down below the beach rocks.
(D) Beneath the torn blanket.

Answer:

- Ⓑ Eyes ablaze, she faced her tormentors.

---

8 Ⓐ Staying at the beach in the hotel with the pool.

Ⓑ Along the highway a lot of flowers blooming.

Ⓒ Everywhere we looked there were beautiful flowers.

Ⓓ Walking for our health and vitality with our friends.

---

9 Ⓐ Swimming during the day and playing games with friends at night.

Ⓑ Want to do the correct homework and get the answers written down.

Ⓒ Children with pretty ribbons and lovely dresses.

Ⓓ I wish I could have a dress just like the one that little girl is wearing.

---

10 Ⓐ Cleaning house on a cold January day.

Ⓑ Eating popcorn and watching movies is just what I like to do.

Ⓒ Other boys and girls playing soccer at the park.

Ⓓ Soccer teams with players ranging from big to small.

Directions: Use this passage and the underlined words to answer questions 11–15.

<u>My friends and me</u> <u>was walking</u> to school
          11   12
one day. <u>We noticed that the house next door</u>
<u>with the windows open</u>. We were worried
              13
because we know that a woman in a wheel-

chair lives in that house. She told me that

her <u>greater</u> fear was that someone would
        14
break into her house. <u>Their</u> had been some
                              15
problems with break-ins in another part of

town.

11 Which is the correct way to write the subject "my friends and me"?

Ⓐ My friends and I
Ⓑ My friends and myself
Ⓒ Me and my friends
Ⓓ As it is

---

12 What is the correct way to write the verb "was walking"?

Ⓐ were walking
Ⓑ is walking
Ⓒ was walk
Ⓓ As it is

---

13 The next sentence says: "We noticed that the house next door with the windows open." Which of the following is needed to make this sentence correct?

Ⓐ We noticed that the house next door with the windows open completely.

Ⓑ We noticed that the house next door had the windows open.

Ⓒ We noticed the house next door that the windows open.

Ⓓ No changes are needed. The sentence is correct as written.

---

14 Which of the following is the correct way to write the word "greater"?

   Ⓐ   greatest
   Ⓑ   more great
   Ⓒ   most great
   Ⓓ   As it is

15 Which of the following is the correct way to write the word "Their"?

   Ⓐ   They're
   Ⓑ   There
   Ⓒ   They
   Ⓓ   As it is

Directions: For questions 16–19, find the part that is the simple subject of the sentence.

Example:

   Sally went to the store to get some milk.
    Ⓐ   Ⓑ   Ⓒ   Ⓓ

Answer:

   Ⓐ   "Sally" is the simple subject.

16 His family goes to the football game
    Ⓐ   Ⓑ   Ⓒ   Ⓓ
   every Saturday in September.

17 Even on rainy days, the group of people
     Ⓐ   Ⓑ   Ⓒ   Ⓓ
   who gather there have a good time.

18 What kind of games do you like to play
       Ⓐ   Ⓑ   Ⓒ
   on Sunday afternoons?
             Ⓓ

19 Mary Ellen is making everyone a
    Ⓐ   Ⓑ
   beautiful sweatshirt to wear.
    Ⓒ   Ⓓ

Directions: For questions 20–22, find the part that is the simple predicate of the sentence.

Example:

   Sally went to the store to get some milk.
    Ⓐ   Ⓑ   Ⓒ   Ⓓ

Answer:

   Ⓑ   "went" is the simple predicate.

20 Jane and her friends always go to the
    Ⓐ   Ⓑ   Ⓒ
   store on Saturday.
    Ⓓ

21 What interests you about football?
    Ⓐ   Ⓑ   Ⓒ   Ⓓ

22 The team of students looked everywhere
          Ⓐ   Ⓑ   Ⓒ
   for the lost ring.
       Ⓓ

Directions: For questions 23 and 24, choose the sentence that best combines the sentences given.

Example:

   Kara went to school.

   She will get there in 15 minutes.

   Ⓐ   Kara went to school and will get there in 15 minutes.

(B) Kara will get to school in 15 minutes.

(C) Kara, who went to school, will arrive in 15 minutes.

(D) Kara went to school and she will get there in 15 minutes.

Answer:

(B) Kara will get to school in 15 minutes.

---

23 Mary Alex loves the beach in the summer.

Katie Beth loves the mountains in the summer.

(A) Mary Alex and Katie Beth love the beach and the mountains in the summer.

(B) Mary Alex loves the beach in the summer, but Katie Beth loves the mountains.

(C) Mary Alex loves the beach in the summer, and Katie Beth loves the mountains in the summer.

(D) Because Mary Alex loves the beach in the summer, Katie Beth loves the mountains.

---

24 Dru will arrive at the airport.

He will arrive in less than one hour.

He will go straight to the hotel.

(A) Dru will arrive at the airport in less than one hour and go straight to the hotel.

(B) Arriving at the airport, Dru will be there in less than an hour and go straight to the hotel.

(C) Since he has to go straight to the hotel, Dru will arrive at the airport in less than an hour.

(D) Dru will arrive at the airport, he will arrive in less than one hour, and he will go straight to the hotel.

Directions: Read the paragraphs below. Find the best topic sentences for the paragraphs.

25 _____. Some choose lake living so they can fish at any time. Others like the view and the peacefulness they feel when gazing out on the water. Still others need the daily enjoyment of swimming or skiing to their heart's content.

(A) Fishing is a fun hobby enjoyed by people of all ages.

(B) Buying a house is a big decision.

(C) Water skiing takes many hours of practice, and it is helpful to live near a lake if you want to develop your skills to the fullest.

(D) People choose to live near a lake for many reasons.

---

26 _____. Many schools have basketball teams open to sixth graders, and some students try out for them. When teams are not available, students usually choose service clubs or interest groups. In addition, some schools offer orchestra and band to students of this age.

(A) Orchestra class is a fun way to spend your free time in sixth grade.

(B) Sixth grade is a time when schools begin to offer a variety of extracurricular activities for students.

(C) The basketball team at our school won all but two of its games last year, and the star of the team was a sixth grader.

(D) The middle school curriculum is very demanding, and students have to study hard to do well.

Directions: Use the paragraph below to answer questions 27–29.

(Sentence 1) Martin has an invitation from his best friend to join him for a basketball game this evening. (Sentence 2) His friend is from California and his name is Allen. (Sentence 3) Martin has a big test tomorrow and he is not sure if he will have time to go to the game. (Sentence 4) He really wants to go though, because he loves basketball and another friend is on the team.

27 Which would be the best first sentence for this paragraph?

(A) Martin has a decision to make in the next few minutes.

(B) Martin is a successful student who is doing well in school.

(C) Basketball is a fun sport because there is so much action.

(D) Martin and his friends always spend a lot of time together.

28 Which sentence is not really needed in the paragraph?

(A) Number one
(B) Number two
(C) Number three
(D) Number four

29 Which sentence would be the best ending one for the paragraph?

(A) Martin will have to think hard about this before he is sure what to do.

(B) Martin will probably go play football in a few minutes.

(C) He is a good friend and a good student.

(D) Martin is not sure how to study for the test and needs some help.

(See page 113 for answer key.)

CHAPTER 6

Spelling and Study Skills

Spelling: As a child, you either loved it or hated it. Lately it seems as if teachers are spending less and less time on the subject, especially by the sixth grade. Most educators believe that spelling is not the most important skill taught by schools. In the earlier grades, children are given spelling lists and students are tested on the words at the end of the week. The concern of parents and educators alike is that memorization of a weekly spelling list often doesn't lead to long-term learning of the words.

What Your Sixth Grader Should Be Learning

By sixth grade, many students don't have regular spelling tests. Rather, they are expected to use spelling skills in writing and reading. They are frequently asked to learn to define words associated with reading assignments. Sometimes, learning to spell the words is also a part of this assignment, but this isn't always the case.

What Tests May Ask

Because of the decreased emphasis on spelling in sixth grade, many standardized tests at this level do not include an entire subtest devoted to spelling skills. Instead, spelling is assessed by subtests that measure written expression, through either multiple-choice or open-ended items.

In either case, if your sixth grader is not a good speller, you can breathe a little easier. It's probably the area that receives the least attention on testing at this level. However, because some tests do measure spelling, we include a discussion of those skills.

Achievement tests usually test spelling by the recognition method. This means that students have to recognize the correct spelling when they see it among other choices. This isn't the way spelling is usually tested in school classes. Rather, students usually have to write down words that their teacher dictates to them.

It's quite a different skill to choose the correct spellings from a list. Some students find it confusing because they are fooled by seeing spellings that look correct. Others find it much easier to demonstrate learning this way because they don't have to generate the spellings themselves.

What You and Your Child Can Do

When practicing spelling with your child, it's important to practice the same way the items will be presented on a test. You may have learned this from experience when trying to practice for classroom spelling tests.

Let's say that you've practiced for a spelling test with your child by calling the words out and having your child spell the words back to you. She spells every word correctly. But when the test comes up the next day, she misses several items. The problem could be that the practice didn't match the testing format. If the teacher calls the words out like in a spelling bee, this method of practice is suitable; but it's not the standard testing method for classroom spelling tests.

The best way to practice for a test *is for the child to demonstrate the skill in the same way it will be tested*. What does this mean for standardized testing? It means that your child needs practice recognizing the correct spelling of words from a list.

Editing As Practice

Practice in recognizing correctly spelled words usually comes from editing someone's work. By examining another's writing, the student has a chance to check for errors of all types and to look for incorrect spellings. To practice this, your child can:

• Reread her own work.

• Trade papers with someone else and read that person's work.

• Read work that has intentional errors in it.

Word Components

Students' spelling usually improves with the learning of word patterns and components. Beginnings, endings, and roots provide good hints for how to spell unfamiliar words.

Many schools today teach word roots and stems as part of the curriculum beginning in sixth grade. Students who learn stems are able to approach new words by breaking them down into word parts.

If you are interested in working with your child on word roots and stems, you can generate a partial list of these stems just by thinking of word parts you know, such as *un-, re-, tri-, bio-* and *hydro-*. However, be aware that sixth graders often don't respond well to straight

memory work. Instead, try these suggestions:

• Point out stems as they appear in words you encounter.

• Notice stems in newspapers, magazines, or books.

• Discuss the similarities and differences in spellings of words we use and how these features may originate with the root word.

How can you get your sixth grader to do this? Try some easy games for the road. Use time in the car for challenging games that focus on developing observational and spelling skills.

• Choose a long word and challenge each person in the car to find the letters of that word in sequence on billboards and signs. If two letters are found on the same billboard, the letters have to be in the same order they are found in the word.

• Adapt the game by finding the letters of a destination name in signs. Choose a destination with a long name—Vancouver, British Columbia is a good choice!

• Watch signs for unusual spellings; give points to each child who can find an unusual spelling (for example, Bar-B-Q, Slush-E's, or Toys R Us). Challenge your child to spell the word in the usual way.

Practice Skill: Spelling

Directions: Follow the directions for each group of questions. Choose the best answer for each question. Be sure to read carefully so that you know whether to look for the word that is spelled correctly or incorrectly.

Directions: For questions 1–3, find the word that is spelled correctly and fits best in the blank.

Example:

The two _____ are erupting!

- (A) volcanos
- (B) volcanoes
- (C) volcanoe
- (D) volcannoes

Answer:

- (B) The two volcanoes are erupting!

---

1 Proceed with _____ when driving in the rain.

- (A) caushion
- (B) caution
- (C) causion
- (D) causchion

---

2 The snow does not _____ to be melting very quickly.

- (A) appear
- (B) apear
- (C) appere
- (D) apeir

---

3 My mother likes _____ candy.

- (A) choclate
- (B) chocalate
- (C) chocolate
- (D) choklit

Directions: For questions 4–7, choose the phrase in which the underlined word is spelled *incorrectly*.

Example:

- (A) make a <u>sugestion</u>
- (B) try a <u>substitute</u>
- (C) <u>discuss</u> the problem
- (D) <u>immense</u> distance

Answer:

- (A) "sugestion" is spelled incorrectly. The correct spelling is "suggestion."

---

4 (A) <u>nether</u> of us
- (B) <u>sneak</u> around
- (C) <u>jealous</u> of the other girls
- (D) sheer <u>delight</u> at the gift

---

5 (A) a <u>colony</u> of bees
- (B) a broken <u>limb</u>
- (C) the single <u>correct</u> answer
- (D) <u>disapointed</u> in her grade

---

6 (A) an <u>attractive</u> girl
- (B) <u>really</u> late
- (C) a <u>radicle</u> change
- (D) <u>celebrate</u> your birthday

---

7 (A) an <u>eagle</u> flying
- (B) <u>hauks</u> in the tree
- (C) a <u>cancellation</u> notice
- (D) the <u>telephone</u> bill

Directions: For questions 8–10, read the phrase. Choose the answer in which the underlined word is spelled incorrectly for the way it is used.

Example:

- (A) the judge is in <u>court</u>
- (B) the king picked his <u>air</u>
- (C) the <u>whole</u> thing
- (D) the <u>pair</u> of shoes

Answer:

- (B) the king picked his <u>air</u>; the correct answer is "heir."

---

8 Ⓐ the <u>seen</u> in the movie
   Ⓑ <u>hear</u> the sound
   Ⓒ the <u>hole</u> in the floor
   Ⓓ the <u>road</u> to town

9 Ⓐ <u>stealing</u> from others
   Ⓑ <u>righting</u> your homework
   Ⓒ the cat's <u>claws</u>
   Ⓓ the <u>latter</u> of the two

10 Ⓐ <u>bare</u> skin
   Ⓑ <u>air</u> of the fortune
   Ⓒ <u>knights</u> of the round table
   Ⓓ the <u>hair</u> on your head

Directions: For questions 11–13, find the underlined word that is misspelled. If all words are spelled correctly, choose D, No mistake.

Example:

I <u>urge</u> you to <u>sign</u> the <u>petetion.</u>
   Ⓐ   Ⓑ   Ⓒ

No mistake.
   Ⓓ

Answer:

   Ⓒ The word should be spelled as "petition."

11 I am looking at the <u>section</u> of seats
                    Ⓐ

<u>where</u> they <u>usually</u> sit. No mistake.
  Ⓑ   Ⓒ   Ⓓ

12 Season tickets are <u>difficult</u> to <u>purchas</u> if
                  Ⓐ   Ⓑ

you are a <u>new</u> fan. No mistake.
      Ⓒ   Ⓓ

13 Holiday <u>traditions</u> are <u>important</u> to
         Ⓐ   Ⓑ

<u>manetain</u> from year to year. No mistake.
  Ⓒ   Ⓓ

(See page 113 for answer key.)

Study Skills

Our society is full of information and information sources. There have never been so many places in which to find and gather data. Conducting research and making the most of study time are important skills for students to learn.

What Your Sixth Grader Should Be Learning

By the sixth grade, students should be able to understand a story web. They should be able to identify and use a wide range of reference sources, including telephone directories, card catalogs (including electronic catalogs online), encyclopedias, and a dictionary. They should be able to utilize various aids within resource books, such as indexes and tables of contents.

Specific skills include the ability to:

- alphabetize words,
- identify organizational methods, and
- use a reference table.

What You and Your Child Can Do

Study skills aren't learned by reading textbooks or reviewing for tests. They are learned through opportunities to search for and work with information sources. The good thing about these skills is that you can have a lot of fun and not realize that you are reviewing for a standardized test!

Let's start with the very first step in preparing to conduct research: developing a plan. Your child has probably had some exposure to this at school. It is best to maintain an awareness of what organizational strategies your child is learning at school so that you can give consistent instructions and supplement those if necessary. Basic skills include:

1. Developing graphic organizers, including a story web

2. Developing an outline for a paper

1. Graphic organizers are visual representations of a story's main points and details that give the writer a structure for how to write a paper. A story web is one type of graphic organizer. Story webs begin with a central idea in a circle. Let's look at one:

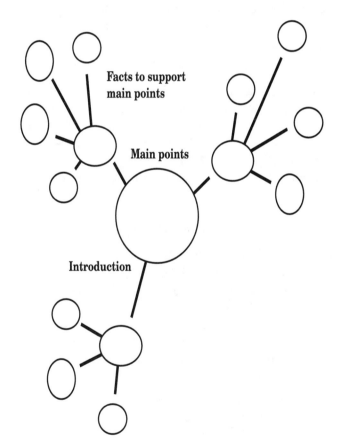

This web starts with a central idea and has three supporting paragraphs. The details for each paragraph are the final bubbles coming off

the second set of bubbles. The student could produce a paper with an introduction and three supporting paragraphs.

Remember, however, that not all students will be able to use graphic organizers successfully. While some students work well with this graphic strategy, others do better with verbal systems of organization. You'll probably be able to tell which style is easier for your child.

2. Outlining is another strategy for organizing material. Sixth graders should have had multiple exposures to outlines, but they may not know how to generate one on their own. An outline format looks something like this:

I. Introduction
   A. Fact
   B. Fact
   C. Fact

II. First paragraph of main ideas
   A. Fact
   B. Fact
   C. Fact

III. Second paragraph of main ideas
   A. Fact
   B. Fact
   C. Fact

IV. Conclusion
   A. Fact
   B. Fact
   C. Fact

Have your child begin with the most important ideas; fill these ideas into the Roman numerals. Then develop secondary ideas and supporting details, and fill these into the spaces after the capital letters. The outline shown here would produce a paper with an introduction, two supporting paragraphs, and a conclusion.

Using Reference Materials

There are many ways to interest your sixth grader in delving into library resources in search of information. But there are also many types of information resources available at home.

Telephone Books. How many times do we have to use the telephone to find items or services? Sixth graders are old enough to help with this task and can probably do a very good job with it. They may be motivated to help because it involves two activities that intrigue them: telephoning and shopping! Turn these tasks over to your child for a few minutes and see if she doesn't learn something from searching through the phone book. Teach the use of various sections of the phone book, including the Yellow Pages, the residential listings, and the government listings. Allow your child to use the telephone book to contact state agencies or local government to answer questions about zoning laws, voting dates, and so on.

Cookbooks. Ask your child to help you find a recipe for a dish you want to cook for a family celebration. Give her a few cookbooks and ask her to check the index and table of contents for recipes for a birthday cake or lasagne planned for her birthday.

Instruction Manuals. After you finish unpacking that new computer, have your sixth grader help you look through all the user manuals that come with it. Check the index and table of contents as the first ways to find out more information.

Catalogs. Look through catalogs for possible birthday or holiday gifts. If your child has been asking for a new bathing suit, allow her to pull out a few catalogs and see how many different kinds she can find. She can practice math skills at the same time by comparing prices and discounts.

Magazines. Allow your child to subscribe to her favorite magazine. Look through it with her when it first arrives and discuss the articles relating to her favorite activities. Note how the table of contents points her in the right direction to look for these.

Dictionaries. Encourage your child to use the dictionary to help define and pronounce words. Let her see you using the dictionary as you find out information about a word you're interested in.

Word Games. Play the "dictionary game" at the dinner table: Have each member of your family come to the table with a new word they've discovered in the dictionary. Have everyone guess the meaning.

Practice Skill: Study Skills

Directions: Read the directions for each section. Choose the answer you think is correct.

Use this part of a telephone directory to answer questions 14 and 15.

Samuels, Sharon	12 Efird St.	989-6723
Sanchez, S R	3 Highgate	234-1151
Sanders, Robert	34 Hope Ferry Rd	767-5546
Scott, F J	68 Dogwood Ln	789-4435
Snead, A R	78 Efird St	345-8897
Stapp, T L	45 Augusta Hwy	892-0909
Sullivan, W J	57 Long Creek Way	898-7765

14 Where does Robert Sanders live?

    Ⓐ 4 Hope Ferry Rd

    Ⓑ 45 Augusta Hwy

    Ⓒ 34 Hope Ferry Rd

    Ⓓ 12 Efird St

15 Sharon Samuels lives very near:

    Ⓐ A R Snead

    Ⓑ T L Stapp

    Ⓒ W J Sullivan

    Ⓓ S R Sanchez

Directions: Use this information from a library card catalog to answer questions 16 and 17.

621.5

TARA LIPINSKI

932
T645

Snow, Sally
Skater with a Dream / by Sally Snow. Photographs by The Sports Photographer. New York: Sports publishers, 1999. 145 p.; photos; 22 cm

1. Lipinski, Tara 2. Athletes, American 3. Athletes, female I. Title

16 What is the title of this book?

Ⓐ Tara Lipinski
Ⓑ Skater with a Dream
Ⓒ Snow, Sally
Ⓓ American Athletes

17 In what section of the card catalog would this card be found?

Ⓐ Title
Ⓑ Subject
Ⓒ Publisher
Ⓓ Author

Directions: Look at the guide words from a dictionary page illustrated here and answer question 18.

snake dance	**sneaky**

18 Which word could be found on the dictionary page illustrated?

Ⓐ smutty
Ⓑ snake
Ⓒ snail
Ⓓ snappy

Directions: Choose the correct answer for the following question:

Example:

Which of these could be a main heading that would include the other three topics?

Ⓐ diabetes
Ⓑ heart disease
Ⓒ health
Ⓓ chicken pox

Answer:

Ⓒ health

19 Which of these could be a main heading that would include the other three topics?

Ⓐ France
Ⓑ Europe
Ⓒ Germany
Ⓓ Spain

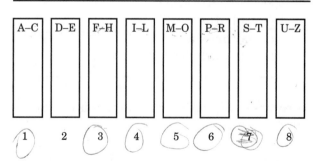

Directions: Look at the encyclopedia volumes illustrated. Each volume contains information about topics that begin with the letters shown on the illustration. Use the information presented to answer questions 20 and 21.

Example:

Which volume would you use to find information about different breeds of horses?

(A) Volume 4
(B) Volume 7
(C) Volume 3
(D) Volume 2

Answer:

(C) Volume 3

20 Which volume would you look at to find information on the rules of tennis?

(A) Volume 7
(B) Volume 6
(C) Volume 1
(D) Volume 5

21 What volume would have information about maples, oaks, and pines?

(A) Volume 5
(B) Volume 6
(C) Volume 7
(D) Volume 8

22 In what reference book would you look for a map of Germany?

(A) a dictionary
(B) an atlas
(C) a thesaurus
(D) an anthology

Directions: Use this outline to answer questions 23 and 24.

My Favorite Sport
I.   Gymnastics
    A. Fun to watch
    B. Fun to participate
    C. Challenging

II. Famous gymnasts
    A. Nadia Comenici
    B. Dominique Moceaneau
    C. Kerry Strug
III. Gymnastic events
    A. Beam
    B. Bars
    C. Floor
    D. Vault
IV. Individual challenge
    A. Set your own goals
    B. Enjoy watching or participating

Example:

Which of these would go best under Roman numeral III?

(A) Mary Lou Retton
(B) Rings
(C) Bela Karolyi
(D) Denver, Colorado

Answer:

(B) Rings

23 Which of these would go best under Roman Numeral II?

(A) The American Olympic gymnastics team
(B) Physical benefits of the sport
(C) Shannon Miller
(D) Olympic coaches

24 If we wanted to add another paragraph to this paper, which would be best?

(A) The benefits of soccer
(B) Weight training
(C) Physical training necessary for gymnastics
(D) Women in sports

(See pages 113–114 for answer key.)

CHAPTER 7

Math Concepts

By the time students get to sixth grade, they have had many math courses that have given them a solid understanding of most math concepts. You may find, however, that your sixth grader has also developed some notions about math that may not be helpful. Many students (especially girls) find math boring or think that the skills they learn aren't relevant. Nothing could be further from the truth! Students who really enjoy math find ways to use it in many aspects of daily life. It will make your job as a parent much easier if you can find ways to make math real and fun for your child.

What Your Sixth Grader Should Be Learning

Sixth graders are expected to have mastered a variety of math concepts, including understanding number systems, number concepts, properties of numbers, and fractions and decimals. Specifically, sixth graders are expected to know a great deal about fractions, prime numbers, factoring, regrouping and rounding, converting and estimating, and using mental math.

What Tests May Ask

Standardized tests of math concepts typically ask a number of questions on a wide variety of math skills, including:

- fractions (ordering fractions and decimals, identifying the lowest common denominator, equivalent fractions, renaming and reducing fractions, and using a number line)
- associating numerals and number words
- whole numbers, integers, and Roman numerals
- factoring numbers and finding the greatest common factor
- decimal place value and converting between decimals and fractions
- number sentences, operational symbols, and properties
- estimating
- finding multiples and square roots
- recognizing numeric patterns, prime and composite numbers
- regrouping and rounding

Mental Math

More and more, math texts and teachers require students to think and solve problems in their heads as much as possible. This process is called mental math, and although it may seem extremely difficult, it's a very effective way to increase your child's understanding of numbers. If you practice, it's not very hard, either. By learning to use strategies like rounding and estimating, it becomes easier to do lots of math calculations in your head. Students will be expected to use mental math on standardized tests.

Have your child read a math item and think about the questions the problem asks first, before he does anything. Then have him look at the numbers in the problem. Numbers that end in zero are often good candidates for mental math. Numbers that are close to hundreds or thousands can be rounded to make mental math easier. It's still possible to make errors in calculations, but the logic of mental math can help you find those errors.

What You and Your Child Can Do

In order to help students with math concepts, parents must first understand how to make math meaningful to students. Real-world examples of math applications abound.

Money

Money is probably the most effective medium for teaching math skills at home. Students in sixth grade should certainly be able to count money and make change. They should be learning how to calculate tax, calculate a sale price, and budget their money. They can learn this best if they are given some money to use as they choose. Providing an allowance is one way of doing this; some parents let their children earn money instead, by doing chores or other tasks. The point is that students who manage money are usually learning math skills at the same time. Using money is a good way to practice mental math because the amounts can be rounded to even dollars or other amounts.

Because children tend to be more interested in figuring out best buys and sale prices if it's their own money at stake, consider investing $10 or $20 in a "store card" offered by many bookstores and toy stores. Although it looks like a charge card, it's really a gift certificate in card form. Allow your child to take the card and go shopping with him. Watch how careful he is to get the most for his money! Have him calculate sales prices; how much is he really saving? Have him compare prices, as well. If there is a "two for one" special on video games, have him figure out the per-unit cost and compare that to the others on the shelf. Have him figure the sales tax.

If your child likes to go food shopping, give him a portion of your list and have him play detective: If the family-size jar of peanut butter at 22 ounces costs $4.95, and the 12-ounce jar costs $2.90, which is really the better buy? Have him do the math and let him choose the best-priced item. This works especially well for kid-friendly items that you may not always buy for your family but that could serve as a reward for "doing the math"—snack foods or candy bars, for example.

Sports Teams

Another super way to teach a number of math skills is by following a sports team. Baseball is an ideal choice, because it's full of statistics and it can be fun to learn how to use them. It makes the game more interesting and fills those long summer days with math practice when the school doors are closed. Just think of these few pieces of data from baseball:

- *Batting averages:* A player's batting average is calculated by dividing hits by the number of "at bats."

- *Earned run average:* A pitcher's earned run average is calculated by dividing the number of earned runs scored while he was pitching by the number of innings he pitched, then dividing by 9. This statistic allows teams to rank pitchers by how many runs they allowed per game.

- *Slugging average:* A batter's slugging average is the number of bases the batter advanced on all at bats divided by the number of times he was at bat.

- *Win–loss averages:* Calculate win–loss averages for a favorite baseball team by dividing the number of games won by the total number of games played.

The morning newspaper is full of these data in the sports section every day in the spring, summer, and fall. That's a lot of the year when you can discuss baseball, read numbers, and follow your favorite team!

Everyday Activities

Opportunities arise within each day to practice math skills and reinforce conceptual understanding. Let's listen in as Bryan and his father discuss their upcoming hiking trip.

DAD: What supplies do you think we'll need?

BRYAN: I've made a list.

DAD: Let's see. How much money do you think we'll need?

BRYAN: Let me estimate how much each of these things costs and I'll add them up and get back to you.

Bryan is motivated to go camping and his father realizes that this is a time when he can gain Bryan's cooperation to get in some math practice. If Dad plays his cards right, Bryan may never know that he's actually doing math and reviewing for a standardized test.

Numeration

Numeration is the study of systems of numbers. Students should understand the concept of quantity and how this relates to the way we write numerals. Early on, of course, students need to understand whole numbers, but eventually they will learn the meanings of simple fractions.

What Your Sixth Grader Should Be Learning

By sixth grade, students are expected to understand more complicated numbers and be able to use different types of numbers, including integers, fractions, mixed numbers, decimals, and Roman numerals. They should also know how to transfer back and forth between decimals and fractions.

The task of ordering numerals follows closely and logically from the task of understanding numeration, quantity, and numerals. When working on this skill, students should understand the amount of a quantity illustrated by a numeral and should put different kinds of numerals in order.

Sixth graders should also be able to use a number line and show where various numerals fall on that line.

What You and Your Child Can Do

Children practice this skill by looking at items together and comparing quantity, thinking about items in pieces or fractions, and reading various kinds of numbers, including decimals and fractions. Here are some tips for everyday activities that help teach these skills:

- Choose a recipe that needs to be adjusted for the number of servings. Tell your child the number of servings needed and have him make the calculations for you.

- Practice reading bigger numbers. Comparing salaries of various professions may give you an opportunity to do this. Students love this because they get to dream about jobs they will have. Have your student look up average salaries of some professions, write them in order, and read them to you.

- Take your child with you on the search for a new car. Sixth graders are very interested in cars and will enjoy comparing prices. When walking around the lot, have your child read the entire number on a vehicle sticker. Make sure he can read it correctly.

- Have your child develop a list of the cars you've looked at in order from highest to lowest price. Calculate the differences in prices. Ask your child to put the information in a table to help you decide which car to purchase.

- On a family trip, share the map with your child and have him calculate distances between points and the shortest way to get to your destination.

- Compare distances traveled on a certain trip to the distance traveled to school each day, to a shopping mall, or on a cross-country expedition.

- Let your child read through a catalog and select items to purchase. Have him calculate the total bill and tax for the items selected, and compare these prices to those available in local stores.

- Assign your student the task of comparison shopping for grocery items using the sale brochures that come with the daily paper.

- Provide lots of opportunities to talk about and read numbers from books, magazines, and newspapers.

- Have your child figure the gas mileage your car is getting. Keep a record of the miles per gallon and compare changes over time.

- Read ingredient lists from cereal boxes and determine which ingredients compose most of the cereal.

- Have your child keep track of what he eats for a day or two. Use a book on calories and food, and have him figure out how many calories he's consumed.

- Ask your child to tell you what fractions are equivalent to the percentages he sees in newspaper ads.

- Have your child look at win–loss percentages for sports teams and calculate them from the number of total games played.

- Ask your child to follow the stock report and note which stocks have increased and by how much. Compare which fractions are greater and which stocks have increased the most or decreased the most.

- Use a number line to illustrate topics of interest to your student, such as records of the school team, your child's averages in various subjects, and so on.

- Discuss the concept of negative numbers as it relates to losses in money when someone buys something and sells it for a lower price. For example, a homeowner buys a house for $100,000 and sells it for $90,000 because he is transferred and has to sell quickly. How much of a loss did he sustain? In contrast, imagine that he buys the same house for $100,000 and sells it for $110,000 because someone really wants the house. How much of a difference is there between the prices? Where would the prices fall on the number line?

Practice Skill: Numeration

Directions: Read and work each problem. Find the correct answer.

Example:

Which group of numbers is ordered from greatest to least?

Ⓐ 3958, 3945, 3992, 3950
Ⓑ 1078, 1125, 1143, 1211
Ⓒ 9678, 9660, 9650, 9543
Ⓓ 3958, 5967, 3048, 3968

Answer:

Ⓒ 9678, 9660, 9650, 9543

1 Which group of numbers is ordered from greatest to least?

Ⓐ 1278, 7128, 2178, 8127
Ⓑ 6511, 5611, 1615, 1165
Ⓒ 2675, 5267, 6257, 7652
Ⓓ 1011, 1101, 11, 1111

2 Which of these groups of numbers is arranged correctly from least to greatest?

Ⓐ .0056, .056, .65, .56
Ⓑ .1212, .212, .210, .2112
Ⓒ .002, .02, .2, 2
Ⓓ .003, .300, .030, 3.00

3 Which of these is expressed by $(8 \times 10{,}000) + (6 \times 1000) + (1 \times 100) + (4 \times 10) + (5 \times 1)$?

- Ⓐ 68154
- Ⓑ 86145
- Ⓒ 8614.5
- Ⓓ 88145

4 Which is another way to write $7 \times 7 \times 7 \times 7$?

- Ⓐ 7×4
- Ⓑ 7 to the fourth power
- Ⓒ 7777
- Ⓓ 4 to the seventh power

5 Which numeral comes right before XX?

- Ⓐ XIV
- Ⓑ XI
- Ⓒ XIIIIIIIII
- Ⓓ XIX

6 Susan wants to have a party for her friends. She is going to serve cupcakes and soft drinks. Cupcakes come in a package of four each, and soft drinks come in a package of six each. What is the smallest number of cupcake packages she must buy in order to have equal numbers of cupcakes and soft drinks?

- Ⓐ 4
- Ⓑ 2
- Ⓒ 6
- Ⓓ 3

7 A teacher wants to store textbooks in boxes for the summer. She finds that she has 260 textbooks. Each box will hold 60 books. How many boxes will she need to pack all the textbooks?

- Ⓐ 5
- Ⓑ 4
- Ⓒ 6
- Ⓓ 3

8 What is the greatest common factor of 81 and 63?

- Ⓐ 7
- Ⓑ 3
- Ⓒ 9
- Ⓓ 27

(See page 114 for answer key.)

Number Concepts

Sixth graders have had practice on a variety of number concepts over the years, but they should review place value of integers (especially in larger numbers), place value of decimals, and even and odd numbers.

Finding Factors; Primes and Composite Numbers

This is a new and important skill for many sixth graders and can be a difficult one to reinforce. Sixth graders are expected to understand how to find the factors of numbers and how to recognize which numbers are prime numbers. A prime number has as factors only itself and one; a composite number has other numbers as its factors. For example, 11 and 17 are prime numbers; 12 and 15 are composite numbers (factors of 12 are 4 and 3 or 2, 2, and 3; factors of 15 are 3 and 5).

> The ability to produce factors for numbers is strongly related to multiplication skills. It's important that students who are working on this skill have a ready grasp of basic multiplication facts, or they won't be accurate in producing factors.

What You and Your Child Can Do

What activities require us to produce factors in daily life? It may take a bit of thinking, but the examples are all around us.

- A family is assigning equal amounts of candy to party bags for birthday party guests. Each bag has 16 pieces of candy. How many different ways can the candy be used to fill goody bags?

- Possible solutions:

  The candy can be broken down into 4 groups of 4 each.

  The candy can be broken down into 8 groups of 2 each.

  The candy can be broken down into 16 groups of 1 each.

Obviously, the family will have to decide how many bags they need to fill in all and how many pieces of candy they will have to put in each.

- If your child wants to make necklaces to sell, she'll need to figure out how many beads will make a certain number of necklaces. If she has 100 beads, how many different ways can she divide up her beads to make necklaces?

  She can make 10 necklaces of 10 beads each.

  She can make 25 necklaces of 4 beads each.

  She can make 4 necklaces of 25 beads each.

  She can make 2 necklaces of 50 beads each.

What she has actually done is produce a set of factors of 100.

Practice Skill: Number Concepts

Directions: Choose the correct answer for the following problems.

Example:

Which number is 1000 less than 795,031?

Ⓐ 895,031
Ⓑ 794,031
Ⓒ 796,031
Ⓓ 784,031

Answer:

Ⓑ 794,031

9 Which number is 1000 less than 678,092?

Ⓐ 668,092
Ⓑ 677,092
Ⓒ 568,092
Ⓓ 68,092

10 Which of the following is a prime number?

Ⓐ 45
Ⓑ 12
Ⓒ 37
Ⓓ 4

11 Which is a composite number?

Ⓐ 11
Ⓑ 17
Ⓒ 15
Ⓓ 5

12 What is the value of the "5" in the numeral 7.405?

Ⓐ 5 hundredths
Ⓑ 5 thousandths
Ⓒ 5 thousand
Ⓓ 5 hundred

(See page 114 for answer key.)

Number Properties

Tests of math properties in sixth grade often measure the student's skill at estimating answers, developing equations for problems, rounding numbers, and finding numbers that make equations true.

Estimating and rounding are two areas where you can help your child succeed. Parents are constantly estimating in order to solve problems, and it's easy and fun to include students in the process.

Mom: I have to bring popcorn to the ballgame tomorrow night. I wonder how many people will be there.

ANNETTE: How can you figure that out?

MOM: Well, I suppose we could guess based on how many people we've seen there before.

ANNETTE: I know! There are about 20 seats on each row.

MOM: Yes, I think you're right. I think there are about 10 rows per section, and there are 3 sections.

ANNETTE: But, how many seats are usually taken?

(And so on ....)

You use rounding numbers all the time. Imagine a shopping trip with a sixth-grade girl for a great instance of using rounding.

CAROLINE: Oh, I love the jeans outfit on display in that window.

MOM: I wonder how much it would cost to buy the whole outfit. Our budget for clothing right now would allow us to spend about $60.

CAROLINE: Well, the blouse is $17.95, the pants are $24.95, and the vest is $12.00. I could get a piece of paper and add it up.

MOM: It'd be faster and easier to round the number and get an estimate. Let's see ... make that 18 plus 25 plus 12 .... Or 25 plus 12 is 37, so round that up to 40 and add 18; that would be 58. Since we rounded up, we know it would actually be a bit less than $58.

Practice Skill: Properties

Directions: Choose the correct answer for the following problems.

Example:

Round 568 to the nearest ten.

- Ⓐ 540
- Ⓑ 550
- Ⓒ 600
- Ⓓ 570

Answer:

- Ⓓ 570

13 Round 347 to the nearest ten.

- Ⓐ 340
- Ⓑ 350
- Ⓒ 300
- Ⓓ 400

14 In which situation would you find estimation a helpful strategy?

- Ⓐ You want to bill your neighbor for the time you spent babysitting.
- Ⓑ You want to buy sodas for the people attending your birthday party.
- Ⓒ You want to tell your parents how many soccer players there are on your team.
- Ⓓ You want to tell your parents how many people were at a football game.

15 What number goes in the blank line to make the sentence true?

$$\_\_\_\_\_ > -6$$

- Ⓐ –3
- Ⓑ –8
- Ⓒ –10
- Ⓓ –12

16 The product of 398×19.87 is closest to:

- Ⓐ 6000
- Ⓑ 800
- Ⓒ 8000
- Ⓓ 10,000

17 Which number completes the number sentence below?

$$7 \times (2 + 5) = \_\_\_\_ + 17$$

- Ⓐ 2
- Ⓑ 32
- Ⓒ 42
- Ⓓ 20

18 One hundred sixth graders were scheduled to go on a field trip to the township auditorium for a performance of one of Shakespeare's plays. On the day of the trip, 10 students were sick and stayed home. Of the remainder, 2 out of 3 reported that they had seen one of Shakespeare's plays performed before. Choose the number sentence that tells how many of the field trip participants had seen a Shakespearean performance before.

Ⓐ $100 - 90 = 2/3 + \_\_\_$
Ⓑ $2/3 \times (100 - 10) = \_\_\_$
Ⓒ $90 + 100 \times 2/3 = \_\_\_$
Ⓓ $(2/3 \times 100) - 10 = \_\_\_$

(See page 114 for answer key.)

Fractions, Decimals, Ratios, and Proportions

Remember the examples of baseball averages we talked about earlier in this chapter? Reread those examples and pick out a team, because this is where you can use them effectively. Put some numbers into those equations for batting averages, slugging averages, and earned-run averages. Open the newspaper and read the statistics for your team. Follow the team and a few players all season.

Students in sixth grade are expected to use ratios and proportions appropriately, to write them in several different ways, and to understand equivalent ratios.

There are many examples in everyday life that call for the use of ratios. Parents can use these examples to increase student's skills:

- Beth needs to purchase apples and oranges in the same proportion a recipe calls for.

- A team member needs to have the right proportion of snacks for the participants in the game on Saturday.

- A parent needs to purchase enough red and blue fabric to make uniforms for team members.

- A student wants to make pancakes with a recipe that calls for one part milk to two parts pancake mix.

- A student wants to make trail mix with different proportions of ingredients so he can decide which proportion is his favorite and make it that way in the future. For example, if your child needs four pounds of mix and wants the mix to have peanuts and raisins, how much of each ingredient will he need to buy if he wants his mix to have equal amounts of raisins and peanuts? How much will he need to buy if he wants his mix to have twice as many peanuts as raisins? The first question asks the student to buy the ingredients in a 1:1 ratio, the second in a 2:1 ratio.

Practice Skill: Fractions and Decimals

Directions: Choose the correct answer for each of the following questions.

Example:

Which of these numbers goes in the blank to make this number sentence true?

$$\_\_\_ > \tfrac{1}{8}$$

Ⓐ $^2/_{16}$
Ⓑ $^4/_{32}$
Ⓒ $^1/_4$
Ⓓ $^1/_{16}$

Answer:

Ⓒ $^1/_4$

19 Which of the following is 6 thousandths?

Ⓐ 0.6000
Ⓑ 0.06
Ⓒ 0.066
Ⓓ 0.006

20 Which fraction is another name for $8^3/_5$?

   Ⓐ $83/_5$
   Ⓑ 8.3
   Ⓒ $43/_5$
   Ⓓ $29/_5$

---

21 Which fraction is in simplest form?

   Ⓐ $10/_{12}$
   Ⓑ $4/_6$
   Ⓒ $3/_5$
   Ⓓ $8/_{12}$

---

22 Which of these numbers goes in the blank to make this number sentence true?

$$1/\underline{\quad} < {}^1/_4$$

   Ⓐ 2
   Ⓑ 3
   Ⓒ 5
   Ⓓ 1

(See page 114 for answer key.)

CHAPTER 8

Math Computation

When most of us think about mathematics, we think about the skills of adding, subtracting, multiplying, and dividing. These are the skills we remember spending hours practicing when we were in school. Not only did we do hundreds of examples of multiplication problems during class, but we also took home page after page of math practice for homework. Those of us who were good at it probably thought the repeated exercises were fun, but some of us struggled to get it just right. Today's math curriculum contains a far wider range of skills than that. This chapter focuses on those good old-fashioned skills of "crunching numbers."

What Your Sixth Grader Should Be Learning

Sixth graders are expected to show mastery of addition, subtraction, division, and multiplication with three groups of numerals: whole numbers, decimals, and fractions. They should pay particular attention to problems that have several steps.

How important are number-crunching skills? You'll get different answers to that question. Some people will tell you that it's essential for all students to be competent at calculating with pencil and paper. Those people will insist that

students have to have memorized the multiplication tables by the time they finish third grade and that if they haven't, they had sure better spend time each night doing just that.

Others will say that there's really no need to spend any time on calculations, since everyone these days uses a calculator or a computer for traditional number-crunching tasks. Why, they say, do students need to waste time on these very intense memory activities when they'll never use them?

What's a parent to do about all of this conflicting advice? Each piece of advice has some merit, and both positions are true to a certain extent. Most educators and most test publishers believe that students need to have it both ways: They need to know how to complete calculations with or without a calculator. There are two reasons why:

1. Students will need to complete calculations when they don't have a calculator around.

2. Students seem to understand the process and the meaning of the process when they have to crunch some numbers themselves.

So, are we saying that it makes sense to spend hours on skill and drill? Are we suggesting that parents give page after page of math practice to prepare for testing? No, of course not. We are saying that students need to have enough experience with computations so they can finish them without a calculator. At the same time, they need experience using a calculator, because without these skills they won't be prepared for the academic and real-life challenges they will face.

Some tests allow students to use calculators; the Scholastic Aptitude Test (SAT) is one. As students get older, the tests they take will measure higher-level skills and will often allow calculator use. Since number-crunching is not the primary focus of such tests, the tests assume that everyone can complete the calculations and they focus on measuring the reasoning skills necessary to set the problem up correctly.

However, most tests for sixth graders don't allow calculators because students at this age are still learning the math processes and need practice on all aspects of math. (There are exceptions to this general rule for students who need specific accommodations because of a disability.)

TIP

Students need practice to use calculators correctly. Some schools provide this practice, because it is a real-life math skill. If your child hasn't had calculator practice at school, make sure you have a calculator at home and allow her to practice with it until she is comfortable.

Before we start talking about math strategies, think for a minute about the children you know. They are very likely quite different in math skills. Some may have learned the calculation skills very easily, while others struggled for long periods of time. If you work with your child a while, you can tell which type of math learner she is. What can you do if your child is struggling to complete the calculations? Try these things:

- Don't get frustrated; that will make the situation worse.
- Don't drill your child so much that she never wants to see another math problem.
- Teach your child to guess at the answer, using rounding and estimation.
- Focus more time on reasoning skills than on calculations; in the long run, it's the more important skill.

Let's get started with strategies that improve computational skills. Remember that the focus of this book is not to provide activities that are just repetitive practice, but to give you ideas about how to build academic practice into real-life experiences.

What You and Your Child Can Do

One of the best strategies you can share with your child is the use of mental math. Not all students are accustomed to using mental math processes, and not all parents are good at it. Mental math allows students to:

- Complete calculations without pencil and paper or calculator.
- Check problems for accuracy without writing them all down.
- Complete math problems quickly, especially when the answer doesn't have to be exact.

Mental math is the process of examining a problem and thinking it through to find a shortcut way to get an answer without using paper. Sometimes the answer will be a guess, based on rounding and estimating. At other times, the answer will be based on regrouping parts of the problem to get round numbers.

Imagine that you need to divide 5128 by 112. To use mental math, round 5128 to 5000 and 112 to 100. Now you can divide two rounded numbers, 5000 by 100, and come up with 50. The answer to your real problem should be around 50.

Addition problems can also be solved using mental math. When adding a column of numbers, such as 8, 4, 2, 6, 7, the mental math way to proceed is first to add the numbers that combine to give a round number. In this case, 8 and 2 make 10 and 6 and 4 make 10. After adding these and getting 20, we only have 7 left to add.

Obviously, there are many other examples of mental math and many different ways to find answers. Engage your sixth grader in this process. It will help her test scores and it may even help her enjoy math more!

Having Fun with Math

Parents and students alike can do lots of fun things with math calculations. Sixth graders have been completing calculations with whole numbers for quite some time, but they may still need practice. They have been adding and subtracting decimals and fractions for a shorter period of time, and will generally find these calculations more challenging.

The use of money is a perfect way to practice. Money is inherently interesting to preadolescents, and it comes in decimals. There are all sorts of real-life ways to practice math with money:

- Determine how many things you can buy for a certain amount.
- Balance a checkbook.
- Create a budget.
- Calculate totals earned from a fund-raiser.
- Figure the tip in a restaurant.

At the Restaurant

DAD: Let's see how much our check is tonight.

CLAUDIA: It's $37.50.

DAD: We need to figure the tip based on that amount. Do you know how to do that?

CLAUDIA: Not really.

DAD: Usually a waiter in a restaurant gets between 15 and 20 percent of the check for a tip.

CLAUDIA: That sounds pretty hard. Let me get out the calculator.

DAD: I'll teach you how to do it quickly. You won't always have a calculator. You know that 10 percent of anything is the amount you get when you move the decimal to the left one place, right? So for our check of $37.50, what would be 10 percent?

CLAUDIA: I guess it would be $3.75, because that's what you get if you move the decimal to the left one place.

DAD: That's right. So to figure 20 percent, what would you do?

CLAUDIA: You would multiply it times 2.

DAD: Yes, you would. That's a little difficult with $3.75. The easiest way from here is to double the dollars, then double the cents. Using that strategy, you would have $6 plus .75 times 2, or 1.50. So, what would a 20 percent tip be?

CLAUDIA: Looks like it would be $7.50.

DAD: Of course. Fifteen percent would be a little less, so you can vary the tip down a little if you didn't think the service was worth a full 20 percent.

Claudia is on her way to becoming a self-reliant young woman, and she's learned some mental math that will carry over to other activities as well. Exactly what skills has Claudia practiced? She multiplied by a fraction: 10 percent or $1/10$. She multiplied a number with a decimal times two, or she added a number with a decimal to itself.

A shopping trip for clothing or school supplies provides another opportunity for improving calculation skills. Some children this age enjoy a little freedom with shopping for these essentials, while others need lots of guidance in their selections. You know what you and your child can handle. Use the shopping trip for a lesson either way. Try these ideas:

1. Give your child a budget and allow her to select items within that budget as long as certain criteria are met. Encourage her to

find sales to help increase her purchasing power. Have her calculate the sale price and add tax so she'll know exactly how much she will spend.

2. Browse through the newspaper and see what sales are advertised that include items your child needs. Allow your child to comparison shop through the newspaper and plan a shopping trip. Calculate sale prices for several shops before you decide where to shop.

Fun with Fractions

Money examples are very useful practice because they contain decimals. Fractions are another matter. They have a terrible reputation and often seem to confuse students; finding practical applications can make the process less intimidating.

Recipes provide one of the best ways to practice fraction skills. Often you need to modify a recipe to serve a different number of people. This is a good opportunity for your sixth grader to practice in a fun context. She may even get to eat the results of the computation!

Let's imagine that a recipe serves 12 people and you want it to serve just the 4 in your family. Turn this task over to your sixth grader and see what happens. She'll need to remember only a few points:

1. You have to divide 12 by 3 to get the correct number to be served.

2. You'll have to divide each ingredient amount by 3 to see how much of that ingredient you'll need for just 4 people.

3. You'll have to make sure you divide each ingredient. Don't forget any!

Other math computation opportunities:

• Play Monopoly with your child and let her be the banker. She'll have to count money, give change, and pay players as they pass "Go."

• Allow your child to help balance your checkbook every now and then. Some parents may be reluctant to give this a try, and it may not be a good idea in every case. But under supervision, sixth graders can do all the operations necessary, and they'll learn some money management skills at the same time. They may even gain some appreciation for the value of money. Have your child do the calculations by hand at first, and then check them with the calculator.

• If your child does well in the checkbook balancing example, consider getting her a checkbook to manage small sums. If local banks won't authorize a checking account for a minor, contact the Young Americans Bank in Denver, Colorado. They provide a range of financial services for even very young children as a way to teach financial responsibility.

• Open a small savings account for your child and allow her to follow the progress of the account by calculating interest, keeping up with the balance, and comparing it to the bank statement.

• Play card games with your child. Many card games, such as Canasta or Gin Rummy, require the players to keep score over several hands. Allow your child to be the scorekeeper and to be responsible for adding the scores after every hand.

• When your child participates in a fund-raiser at school, Scouts, or band, give her some responsibility for keeping up with the number of items she's sold, the amount of money due, the amount of sales tax, if applicable, and so on.

Practice Skill: Addition

Directions: Choose the correct answer to each addition problem. Choose "None of these" if the correct answer is not given.

Example:

$1/4 + 2/4 =$ _____

(A) $3/4$
(B) $1/8$
(C) $4/4$
(D) None of these

Answer:

(A) $3/4$

---

1   50.6
   + 5.3

(A) 505.9
(B) 55.9
(C) 55.09
(D) None of these

---

2   2305
   +6497

(A) 8812
(B) 8892
(C) 8792
(D) None of these

---

3 $1/8 + 2/8 =$ ____

(A) $3/8$
(B) $3/16$
(C) $3/4$
(D) None of these

---

4   653
   429
   + 532

(A) 1514
(B) 1614
(C) 1515
(D) None of these

---

5 $1/5 + 1/10 =$ ____

(A) $1/15$
(B) $2/15$
(C) $3/10$
(D) $1/5$
(E) None of these

Practice Skill: Subtraction

Directions: Choose the correct answer for each question. Choose "NG" if the answer is not given.

Example:

   622
   −63

(A) 778
(B) 775
(C) 559
(D) 659

Answer:

(C) 559

---

6   745
   −67

(A) 778
(B) 678
(C) 688
(D) 698
(E) NG

---

7 $0.504 - 0.007 =$ ____

(A) 0.511
(B) 1.511
(C) 0.498
(D) 0.497
(E) NG

---

8 $7/8 - 1/2 =$ ___

  Ⓐ $1/2$
  Ⓑ $2/8$
  Ⓒ $1/4$
  Ⓓ $6/8$
  Ⓔ NG

9 $8^5/6$
    $-4^2/3$

  Ⓐ $4^1/6$
  Ⓑ $3^3/6$
  Ⓒ $4^3/6$
  Ⓓ $13^1/2$
  Ⓔ NG

Practice Skill: Multiplication

Directions: Choose the correct answer for each question. Choose "NG" if the answer is not given.

Example:

  6.02
  $\times 1.3$

  Ⓐ 7.826
  Ⓑ 78.26
  Ⓒ .7826
  Ⓓ 7826

Answer:

  Ⓐ 7.826

10 7.36
    $\times 2.7$

  Ⓐ 198.72
  Ⓑ 1987.2
  Ⓒ 19.872
  Ⓓ 18.872
  Ⓔ NG

11 ___ $\times 60 = 2100$

  Ⓐ 40
  Ⓑ 100
  Ⓒ 35
  Ⓓ 200
  Ⓔ NG

12 $2/3 \times 15 =$ ___

  Ⓐ $2/45$
  Ⓑ 10
  Ⓒ $10^1/3$
  Ⓓ 9
  Ⓔ NG

13 $.8 \times .7 =$ ___

  Ⓐ 5.6
  Ⓑ 56
  Ⓒ .56
  Ⓓ .056
  Ⓔ NG

Practice Skill: Division

Directions: Choose the correct answer for each question. Choose "NG" if the answer is not given.

Example:

  $46.0 \div 2.3 =$ _____

  Ⓐ 4
  Ⓑ 6
  Ⓒ 20
  Ⓓ NG

Answer:

  Ⓒ 20

14 $12.22 \div 2.6 =$ ___

  Ⓐ 47
  Ⓑ .47
  Ⓒ 4.7
  Ⓓ NG

15 $5/6 \div 1/2 =$ ___

   Ⓐ $5/12$

   Ⓑ $2.5/6$

   Ⓒ $1 2/3$

   Ⓓ 2

   Ⓔ NG

16 $905 \div 36 =$ ___

   Ⓐ 25

   Ⓑ 26

   Ⓒ 25 R 5

   Ⓓ 20 R 5

   Ⓔ NG

Directions: Estimate the answer for this item.

17 $3894 \div 21$ is closest to:

   Ⓐ 100

   Ⓑ 200

   Ⓒ 20

   Ⓓ 10

   Ⓔ NG

(See page 114 for answer key.)

CHAPTER 9

Math Applications

Schools and teachers are spending more time attempting to help students learn to apply mathematics, because the mastery of these skills leads students toward more complicated thinking and learning. Students need to understand these concepts if they are to study higher-level science, engineering, or technology. With today's emphasis on computers and technology, you can see why these skills are among the most important for students.

What Tests May Ask

You will find a heavy reliance on math applications skills in standardized tests. Math applications tests for sixth graders include some complicated items that may challenge parents and students alike, including items measuring competence in geometry, measurement, problem solving, and algebra. Although this is a difficult area for many students, it's one of the areas that provides many opportunities for parents to practice skills with students at home.

What Your Sixth Grader Should Be Learning

Among other things, students in sixth grade may be asked to do a range of very practical things, such as estimate weight and size; use tables, charts, and graphs; and use standard and metric units of measurement. They'll need to know how to read a thermometer, figure out time, and recognize the value of money and money notation. Today's tests emphasize consumer applications, including calculating sale prices, tax, and tips.

Students will also be asked to do traditional math activities, such as solve word problems and formulate a simple number sentence. They must be able to recognize plane and solid figures; find perimeter, area, and volume; and understand points, lines, segments, and angles. Students should understand probability, averages, inequalities, and combinations; be able to identify information needed to solve a problem; be able to solve simple equations; and understand ratio and proportion. Sixth graders should also expect questions on data analysis, statistics, and graphs.

What You and Your Child Can Do

Sixth grade is packed with instruction and testing on math applications, so let's get down to some strategies to bring these concepts home to students. Once you begin to think about ways to practice applications of these skills, you'll come up with hundreds of examples in everyday life.

Let's imagine that a family of four just woke up on a Saturday morning. As the family eats breakfast, sixth grader Mary Ellen wonders what the nutritional value of the cereal is and proceeds to check the label and read the charts. She compares two cereals and notes that one provides 4% of dietary fiber while the other provides 32%. She also notes that the cereals vary in the percentage of Vitamins A, C, and D provided. After doing a thorough review of the ingredients and nutritional analysis, she decides that the first one has more nutritional value. As she examines the boxes, she notices that the smaller box is actually heavier than the larger one. The smaller box weighs 18 ounces (510 grams) while the larger one weighs 12 ounces (340 grams). In addition, there are 9 servings in the smaller box, and 12 in the larger. Her mother tells her that the first cereal costs $3.98 per box and the second costs $4.69 per box. She proceeds to calculate the cost of a serving of each cereal.

Dad turns to the task of planning the day: He wants to go to the hardware store to buy some lumber for the deck he's building. He discusses the dimensions, and all four go outside to measure the space available and calculate the area of the deck before planning their need for supplies. Dan decides to go with Dad to buy the lumber and other supplies. He begins to get ready and starts to take a shower.

When he steps in the shower, he notices that the water isn't very hot; soon all the hot water is gone. He yells out to the other family members after getting out of the cold shower and wonders how many gallons of hot water the family has gone through already that day. Mom reports she has washed a load of clothes and taken a shower; Dad has taken a shower and washed some dishes. Mary Ellen has taken a shower, too. Dan and Dad go to check the size of the hot water heater to determine how much water was available.

It's easy for parents to miss these opportunities to illustrate math concepts. We move through our busy days eager to accomplish the day's chores and get in a little recreation. But to reinforce math, you don't have to talk math constantly. It's not necessary to catch every opportunity because there are so many illustrations of these concepts in daily life. Often parents only need to point out the places where math skills are used.

You don't have to be a math scholar to find ways to introduce your child to everyday math. In all likelihood, the concepts will come back to you as you review with your child. Glance through the math textbook a few times. If you still feel a bit uncomfortable about helping your child with math, try these tips:

- Study your child's math homework and have him explain what he's learning. Your child is probably learning a slightly different math vocabulary and procedure than you learned.

- Meet with your child's math teacher and ask for guidance in how math is taught to your child.

- If you don't understand the skills, consider turning this area over to another family member who's better at math than you are.

- Have your child work together with another student.

- If all else fails, consider hiring a tutor for some sessions with your child. You may want to listen in on the sessions so you can increase your understanding as well.

Geometry

Geometry is the study of figures in space. Sixth graders are expected to recognize and describe geometric shapes and to use geometry vocabulary. This includes the labels of various plane and solid shapes, including angles, rays, planes, cones, cubes, spheres, rectangular prisms, cylinders, and pyramids.

If your child is to do well in geometry, he'll need to understand vocabulary that you don't often use around the house. These words may not occur naturally in our vocabularies, and we need to be aware of them so that our children can have additional practice:

Line: A straight line extending in both directions.

Point: A single point, best illustrated by a dot.

Ray: A straight line that extends in one direction but ends in the other at a point.

Plane: A flat surface that extends in all directions, best illustrated by a wall that goes on forever.

Congruence: When two geometric shapes are of the same size and shape.

Vertex: The point of connection of two rays to form an angle.

Bisector of a line segment: The point at which the line segment is divided into two equal line segments.

Bisector of an angle: A line that passes through the vertex of the angle and divides the angle into two equal angles.

You may have to rely on textbook examples or artificial situations to practice the words listed above. Other words can be included easily in daily conversation. Maps of your town, directions to destinations, or descriptions of objects may include these words, such as intersection, parallel, perpendicular, face, and line segments.

Practice Skill: Geometry

Directions: Find the correct answer for each geometry problem.

Example:

A plane figure with 8 sides is called:

Ⓐ an octagon
Ⓑ a hexagon
Ⓒ a quadrilateral
Ⓓ a pentagon

Answer:

Ⓐ an octagon

1 These two lines are:

Ⓐ similar   Ⓑ parallel
Ⓒ perpendicular   Ⓓ equal

2 Choose the obtuse angle:

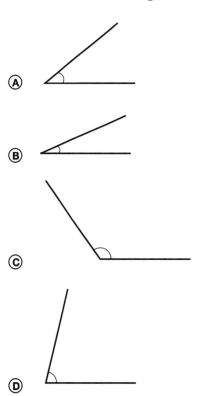

3 Choose the congruent figures from the four listed below:

Ⓐ 1 and 4
Ⓑ 2 and 3
Ⓒ 1 and 3
Ⓓ 1 and 2

4 A plane figure with 5 sides is called a (an):

- Ⓐ octagon
- Ⓑ hexagon
- Ⓒ quadrilateral
- Ⓓ pentagon

(See page 114 for answer key.)

Measurement

Sixth graders are expected to use the skill of applying measurement systems appropriately. Students will be tested on their skills at:

- Using the concepts of length, capacity, weight, perimeter, area, volume, time, and temperature measurement.
- Using metric and standard systems of measurements.
- Estimating and using measurement for description and comparison.
- Using tools for measuring.
- Finding the perimeter and the area of common two-dimensional shapes.

What You and Your Child Can Do

Parents can find many places where students can practice their measurement skills. Try these:

- Have your child keep track of his height and weight in inches and centimeters, pounds and kilograms. Keep a chart and show changes in height and weight over time.
- Shop for groceries together. Label containers by their measurement. We commonly say we want a gallon of milk, but we can also ask for a half-gallon of juice, a liter of soft drink, a cup of yogurt, or a quart of buttermilk.
- Use various examples to ask your child to estimate weight, length, and volume. Ask her to guess how much a book weighs, how long a sofa is, how much a jar contains, and so on. After guessing, check your guesses by measuring the objects.

- Determine the perimeter and area of your house lot. Measure your lot on each side.
- Discuss the relative sizes of containers you purchase. Most containers sold in grocery stores now have both metric and standard measurements written on them. Note the relationships between ounces and grams, liters and quarts.
- Discuss the time elapsed between your activities. Plan ahead by figuring how long you'll be involved in activities and what time you'll return home.
- Keep a thermometer at home and read it each morning before school. Plan clothing and activities based on the temperature.
- When you decide to move furniture around, involve your child by having her measure the space and the furniture involved. Start by estimating the space available and the measurements of the furniture.

Practice Skill: Measurement

Directions: Find the correct answer for each problem.

Example:

A container of herbs weighs:

- Ⓐ a few ounces
- Ⓑ a few pounds
- Ⓒ a few grams
- Ⓓ a few cups

Answer:

- Ⓐ a few ounces

5 A container of french fries weighs:

- Ⓐ a few ounces
- Ⓑ a few pounds
- Ⓒ a few grams
- Ⓓ a few milligrams

6 A bottle of fruit juice contains 1.89 liters of juice. About how many quarts is this?

- (A) 1
- (B) 1½
- (C) 2
- (D) 3

7 A movie starts at 7:30 p.m. and lasts 2 hours and 45 minutes. It takes 30 minutes to get home from the theater. What time should Frankie tell her mother that she'll be home?

- (A) about 10:15 p.m.
- (B) about 10:45 p.m.
- (C) about 11:00 p.m.
- (D) about 11:15 p.m.

8 A sofa is about 75 inches long. What's another way of expressing its length?

- (A) about 5 feet
- (B) a little over 6 feet
- (C) about 3 yards
- (D) about 1 yard

(See page 114 for answer key.)

Problem Solving

Solving word problems is an important application of math skills and one that sixth graders have been practicing for years. It's still a difficult one for many students, so parents should be aware of the need for additional practice.

Problems usually contain key words or phrases telling the student what operation to use. Here are some hints for helping your child solve word problems:

- "In all" usually means add.

- "How much change" usually means subtract.

- "How many more" usually means subtract.

- Students should read problems slowly and think through the steps. Rereading the problem more than once may make it more clear.

- Some students benefit from drawing a diagram of what the problem is describing.

- Some students understand the problem but use the wrong numbers or miscopy the numbers from the problem. Encourage your student to check her numbers before solving the problem.

Standardized tests for sixth graders ask students to interpret graphs and charts. To make sure your child is familiar with both, look for examples of these around the house and use them when opportunities arise.

Newspapers and Magazines. Newspapers and magazines are a good source of graphs and charts (especially the national daily *USA Today*, which is available at most newsstands). Encourage your child to check for graphs and charts and cut them out. Save them and discuss them with your child. Typical topics of graphs include economic forecasts, precipitation levels, and so on. Food labels often contain charts listing nutritional content. Read these regularly with your child.

Listening Game. Let's imagine that a family is on the way to soccer practice and has a few minutes in the car together. This is a good time to practice listening skills and word problems in a fun way. Consider this scenario. There are two children in the car, one in third grade and one in sixth grade. Notice that the mother starts a story and builds in math problems, some approximately on the third-grade level and some approximately on the sixth-grade level. The mother shifts back and forth between the two children as she gives problems. Don't worry if you aren't sure of exactly what level problems your child should have. The problems will always require careful listening, deciding what operation to apply, and remembering the numbers presented. Those skills are important to students of all ages.

MOM: [Starts a story.] There's a family living in town with three children. All three want to play soccer. Caroline [the third grader], how many soccer shoes does this family have to buy for their three children to have one pair each?

CAROLINE: Six!

MOM: Mary [the sixth grader], how many shoes does the family have to buy for each child to have two pairs of shoes and for each adult to have four pairs of shoes?

MARY: Oh! That would be 3 children × 2 pairs of shoes × 2 shoes in a pair plus 2 adults × 8 shoes each. So 3 × 2 × 2 equals 12 and 12 plus 16 equals 28—28 shoes.

MOM: Yes. Now the family needs to go shopping for school clothes. How much will they spend if they buy 5 pairs of jeans at $12 each, Caroline?

CAROLINE: Wow! That would be $12.00 times 5. Let me see if I can figure that out in my head. 12 times 2 is 24, so 12 times 4 is 48. One more 12 would be ... 60.

MOM: Good job. Now, Mary, the family buys the 5 pairs of jeans. They also buy 3 shirts at $11 each and 10 pairs of socks for $6. The mother pays with a $100 bill. If there is no sales tax, how much change would she get back?

MARY: They spent $60 on the jeans. Then they spent 3 × $11 on the shirts; that would be another $33. And $6 for socks. So the total would be $60 plus $33 plus $6. That would be ... $99.

MOM: So how much change did she get back?

MARY: Only $1!

Data Analysis

Sixth graders need to have experience with analyzing data to do well on standardized tests. Analyzing data means looking at a group of scores or numbers and making statements about them. A group of scores on an exam might look something like this:

82 90 82 100 82 60 70 95 85 87 89

To describe the scores, we might use words like:

1. *Mean:* The average of the scores. (In our example, it's 83.8.)

2. *Median:* The middle score, the one that would have half of the scores above it and half below it. (In our example, it's 85.)

3. *Mode:* The most common score. (In our example, it's 82.)

4. *Range:* The lowest to highest scores. (In this example, the range is from 60 to 100.)

Practice Skill: Problem Solving

Directions: Find the correct answer for each question.

Example: The corner grocery sells filet mignon for $12 a pound. On Friday they run a special and lower the price by 10%. Some meat is still left by the end of the day, so to make sure they sell all the meat they reduce the original price by 30%. What was the final price of the steak?

  Ⓐ $8.50
  Ⓑ $8.40
  Ⓒ $10.80
  Ⓓ $3.60

Answer:

  Ⓑ $8.40

---

9 A dress originally sold for $80. The store first put the dress on sale for 25% off regular price. The dress didn't sell, so the store reduced the sale price by 50%. What was the final price of the dress?

  Ⓐ $40
  Ⓑ $45
  Ⓒ $30
  Ⓓ $37.50

10 A cashier at a grocery store rings up an average of 21 customers an hour. How would you determine how many customers she will check out in her 8-hour shift on an average day?

Ⓐ $21 \div 8 =$ _____
Ⓑ $21 \times 8 =$ _____
Ⓒ $21 + 8 =$ _____
Ⓓ $2(21 + 8) =$ _____

Directions: Use the following table to answer question 11.

Train Schedule

Train destinations	Charleston	Aiken	Greenville
Departure times	9:03 a.m.	9:30 a.m.	10:00 a.m.
Arrival times	11:07 a.m.	10:46 a.m.	12:11 p.m.

11 If you caught the train to Aiken, about what time would you need to ask someone to meet you at the station in Aiken?

Ⓐ 11:00 p.m.
Ⓑ 11:00 a.m.
Ⓒ 12:15 p.m.
Ⓓ 10:45 a.m.

Directions: Use the following distribution of scores to answer question 12.

Number of scores: 25

Scores:
96	85	72
78	55	75
70	98	85
65	85	82
66	85	68
55	77	
90	67	
99	99	
78	95	
88	91	

12 What is the range of scores in the distribution?

Ⓐ 65–100
Ⓑ 50–99
Ⓒ 0–100
Ⓓ 55–99

Algebra

Algebra requires the use of equations to solve problems involving missing variables. Standardized tests for sixth graders include items assessing how well your child can solve problems using variables, number sentences, equations, and inequalities.

The most important concept for students to master at this point is that of a variable. Remember that the word *variable* comes from the word *vary,* which means "change." So a variable is a value that changes from problem to problem and changes based on the other values in the equation. A variable is a missing number in an equation. It may be signified by letters that stand for the value. Here are examples of ways equations may be written with a variable:

In the equation $x + 3 = 5$, x is the variable. In this particular problem, it equals 2.

In the problem: $8 +$ ____ $= 11$, the variable is the blank. Its value is 3.

You can help your child with this concept by making up examples of number sentences including variables. For example: Nancy spent $25 on a shirt and a pair of jeans. Her shirt cost $12.95. How much did her jeans cost? You could write that problem as a number sentence this way:

$$x + 12.95 = 25$$
$$\text{OR}$$
$$25 - 12.95 = x$$

x is the value of the jeans.

Consider this problem: $x < 212$ and $x > 156$. What values could x take on? This problem is an example of a situation in which x can have many values, as long as they are less than 212 and greater than 156 (possible values are 157, 158, 159, and so on, up to 211).

Sixth graders also need to remember these key concepts:

- When solving an equation, you can perform the same operation on each side of the equation without changing the value of the equation.

- When solving an equation, it's helpful to use opposing operations to remove information from one side of an equation so that only the variable remains on that side.

Example:

$$Y + 8 = 72$$
$$Y + 8 - 8 = 72 - 8$$
$$Y = 72 - 8$$
$$Y = 64$$

Practice Skill: Algebra

Directions: Choose the correct answer for each question.

Example:

If $X < 68$ and $X > 57$, what could be a value of X?

  Ⓐ 56
  Ⓑ 43
  © 67
  Ⓓ 57

Answer:

  © 67

13 What is the value of X in the equation $10(X + 1) = 100$?

  Ⓐ 9
  Ⓑ 10
  © 8
  Ⓓ 1

14 If $Z < 57$ and $Z > 43$, what could be a value of Z?

  Ⓐ 58
  Ⓑ 43
  © 47
  Ⓓ 57

15 If 12 people arrange to stay in two condominiums at the beach and one condominium has two times as many people as the other one, how many people are in each condominium?

  Ⓐ 6 in each
  Ⓑ 8 in one and 4 in the other
  © 6 in one and 3 in the other
  Ⓓ 9 in one and 3 in the other

(See page 114 for answer key.)

Web Sites and Resources for More Information

Homework

Homework Central
http://www.HomeworkCentral.com
Terrific site for students, parents, and teachers, filled with information, projects, and more.

Win the Homework Wars
(Sylvan Learning Centers)
http://www.educate.com/online/qa_peters.html

Reading and Grammar Help

Born to Read: How to Raise a Reader
http://www.ala.org/alsc/raise_a_reader.html

Guide to Grammar and Writing
http://webster.commnet.edu/hp/pages/darling/grammar.htm
Help with "plague words and phrases," grammar FAQs, sentence parts, punctuation, rules for common usage.

Internet Public Library: Reading Zone
http://www.ipl.org/cgi-bin/youth/youth.out

Keeping Kids Reading and Writing
http://www.tiac.net/users/maryl/

U.S. Dept. of Education: Helping Your Child Learn to Read
http://www.ed.gov/pubs/parents/Reading/index.html

Math Help

Center for Advancement of Learning
http://www.muskingum.edu/%7Ecal/database/Math2.html
Substitution and memory strategies for math.

Center for Advancement of Learning
http://www.muskingum.edu/%7Ecal/database/Math1.html
General tips and suggestions.

Math.com
http://www.math.com
The world of math online.

Math.com
http://www.math.com/student/testprep.html
Get ready for standardized tests.

Math.com: Homework Help in Math
http://www.math.com/students/homework.html

Math.com: Math for Homeschoolers
http://www.math.com/parents/homeschool.html

The Math Forum: Problems and Puzzles
http://forum.swarthmore.edu/library/resource_types/problems_puzzles
Lots of fun math puzzles and problems for grades K through 12.

The Math Forum: Math Tips and Tricks
http://forum.swarthmore.edu/k12/mathtips/mathtips.html

Tips on Testing

Books on Test Preparation
http://www.testbooksonline.com/preHS.asp
This site provides printed resources for parents who wish to help their children prepare for standardized school tests.

Core Knowledge Web Site
http://www.coreknowledge.org/
Site dedicated to providing resources for parents; based on the books of E. D. Hirsch, Jr., who wrote the *What Your X Grader Needs to Know* series.

Family Education Network
http://www.familyeducation.com/article/0,1120, 1-6219,00.html
This report presents some of the arguments against current standardized testing practices in the public schools. The site also provides links to family activities that help kids learn.

Math.com
http://www.math.com/students/testprep.html
Get ready for standardized tests.

Standardized Tests
http://arc.missouri.edu/k12/
K through 12 assessment tools and know-how.

Parents: Testing in Schools

KidSource: Talking to Your Child's Teacher about Standardized Tests
http://www.kidsource.com/kidsource/content2/ talking.assessment.k12.4.html
This site provides basic information to help parents understand their children's test results and provides pointers for how to discuss the results with their children's teachers.

eSCORE.com: State Test and Education Standards
http://www.eSCORE.com
Find out if your child meets the necessary requirements for your local schools. A Web site with experts from Brazelton Institute and Harvard's Project Zero.

Overview of States' Assessment Programs
http://ericae.net/faqs/

Parent Soup
Education Central: Standardized Tests
http://www.parentsoup.com/edcentral/testing
A parent's guide to standardized testing in the schools, written from a parent advocacy standpoint.

National Center for Fair and Open Testing, Inc. (FairTest)
342 Broadway
Cambridge, MA 02139
(617) 864-4810
http://www.fairtest.org

National Parent Information Network
http://npin.org

Publications for Parents from the U.S. Department of Education
http://www.ed.gov/pubs/parents/
An ever-changing list of information for parents available from the U.S. Department of Education.

State of the States Report
http://www.edweek.org/sreports/qc99/states/ indicators/in-intro.htm
A report on testing and achievement in the 50 states.

Testing: General Information

Academic Center for Excellence
http://www.acekids.com

American Association for Higher Education Assessment
http://www.aahe.org/assessment/web.htm

American Educational Research Association (AERA)
http://aera.net
An excellent link to reports on American education, including reports on the controversy over standardized testing.

American Federation of Teachers
555 New Jersey Avenue, NW
Washington, D.C. 20011

Association of Test Publishers Member Products and Services
http://www.testpublishers.org/memserv.htm

Education Week on the Web
http://www.edweek.org

ERIC Clearinghouse on Assessment and Evaluation
1131 Shriver Lab
University of Maryland
College Park, MD 20742
http://ericae.net
A clearinghouse of information on assessment and education reform.

FairTest: The National Center for Fair and Open Testing
http://fairtest.org/facts/ntfact.htm
http://fairtest.org/
The National Center for Fair and Open Testing is an advocacy organization working to end the abuses, misuses, and flaws of standardized testing and to ensure that evaluation of students and workers is fair, open, and educationally sound. This site provides many links to fact sheets, opinion papers, and other sources of information about testing.

National Congress of Parents and Teachers
700 North Rush Street
Chicago, Illinois 60611

National Education Association
1201 16th Street, NW
Washington, DC 20036

National School Boards Association
http://www.nsba.org
A good source for information on all aspects of public education, including standardized testing.

Testing Our Children: A Report Card on State Assessment Systems
http://www.fairtest.org/states/survey.htm
Report of testing practices of the states, with graphical links to the states and a critique of fair testing practices in each state.

Trends in Statewide Student Assessment Programs: A Graphical Summary
http://www.ccsso.org/survey96.html
Results of annual survey of states' departments of public instruction regarding their testing practices.

U.S. Department of Education
http://www.ed.gov/

Web Links for Parents Who Want to Help Their Children Achieve
http://www.liveandlearn.com/learn.html
This page offers many Web links to free and for-sale information and materials for parents who want to help their children do well in school. Titles include such free offerings as the Online Colors Game and questionnaires to determine whether your child is ready for school.

What Should Parents Know about Standardized Testing in the Schools?
http://www.rusd.k12.ca.us/parents/standard.html
An online brochure about standardized testing in the schools, with advice regarding how to become an effective advocate for your child.

Test Publishers Online

ACT: Information for Life's Transitions
http://www.act.org

American Guidance Service, Inc.
http://www.agsnet.com

Ballard & Tighe Publishers
http://www.ballard-tighe.com

Consulting Psychologists Press
http://www.cpp-db.com

CTB McGraw-Hill
http://www.ctb.com

Educational Records Bureau
http://www.erbtest.org/index.html

Educational Testing Service
http://www.ets.org

General Educational Development (GED) Testing Service
http://www.acenet.edu/calec/ged/home.html

Harcourt Brace Educational Measurement
http://www.hbem.com

Piney Mountain Press—A Cyber-Center for Career and Applied Learning
http://www.pineymountain.com

ProEd Publishing
http://www.proedinc.com

Riverside Publishing Company
http://www.hmco.com/hmco/riverside

Stoelting Co.
http://www.stoeltingco.com

Sylvan Learning Systems, Inc.
http://www.educate.com

Touchstone Applied Science Associates, Inc. (TASA)
http://www.tasa.com

Tests Online

(*Note:* We don't endorse tests; some may not have technical documentation. Evaluate the quality of any testing program before making decisions based on its use.)

Edutest, Inc.
http://www.edutest.com
Edutest is an Internet-accessible testing service that offers criterion-referenced tests for elementary school students, based upon the standards for K through 12 learning and achievement in the states of Virginia, California, and Florida.

Virtual Knowledge
http://www.smarterkids.com
This commercial service, which enjoys a formal partnership with Sylvan Learning Centers, offers a line of skills assessments for preschool through grade 9 for use in the classroom or the home. For free online sample tests, see the Virtual Test Center.

Read More about It

Abbamont, Gary W. *Test Smart: Ready-to-Use Test-Taking Strategies and Activities for Grades 5–12. Upper Saddle River,* NJ: Prentice Hall Direct, 1997.

Cookson, Peter W., and Joshua Halberstam. *A Parent's Guide to Standardized Tests in School: How to Improve Your Child's Chances for Success.* New York: Learning Express, 1998.

Frank, Steven, and Stephen Frank. *Test-Taking Secrets: Study Better, Test Smarter, and Get Great Grades (The Backpack Study Series).* Holbrook, MA: Adams Media Corporation, 1998.

Gilbert, Sara Dulaney. *How to Do Your Best on Tests: A Survival Guide.* New York: Beech Tree Books, 1998.

Gruber, Gary. *Dr. Gary Gruber's Essential Guide to Test-Taking for Kids, Grades 3–5.* New York: William Morrow & Co., 1986.

———. *Gary Gruber's Essential Guide to Test-Taking for Kids, Grades 6, 7, 8, 9.* New York: William Morrow & Co., 1997.

Leonhardt, Mary. *99 Ways to Get Kids to Love Reading and 100 Books They'll Love.* New York: Crown, 1997.

———. *Parents Who Love Reading, Kids Who Don't: How It Happens and What You Can Do about It.* New York: Crown, 1995.

McGrath, Barbara B. *The Baseball Counting Book.* Watertown, MA: Charlesbridge, 1999.

———. *More M&M's Brand Chocolate Candies Math.* Watertown, MA: Charlesbridge, 1998.

Mokros, Janice R. *Beyond Facts & Flashcards: Exploring Math with Your Kids.* Portsmouth, NH: Heinemann, 1996.

Romain, Trevor, and Elizabeth Verdick. *True or False?: Tests Stink!* Minneapolis: Free Spirit Publishing Co., 1999.

Schartz, Eugene M. *How to Double Your Child's Grades in School: Build Brilliance and Leadership into Your Child—from Kindergarten to College—in Just 5 Minutes a Day.* New York: Barnes & Noble, 1999.

Taylor, Kathe, and Sherry Walton. *Children at the Center: A Workshop Approach to Standardized Test Preparation, K–8.* Portsmouth, NH: Heinemann, 1998.

Tobia, Sheila. *Overcoming Math Anxiety.* New York: W. W. Norton & Company, Inc., 1995.

Tufariello, Ann Hunt. *Up Your Grades: Proven Strategies for Academic Success.* Lincolnwood, IL: VGM Career Horizons, 1996.

Vorderman, Carol. *How Math Works.* Pleasantville, NY: Reader's Digest Association, Inc., 1996.

Zahler, Kathy A. *50 Simple Things You Can Do to Raise a Child Who Loves to Read.* New York: IDG Books, 1997.

What Your Child's Test Scores Mean

Several weeks or months after your child has taken standardized tests, you will receive a report such as the TerraNova Home Report found in Figures 1 and 2. You will receive similar reports if your child has taken other tests. We briefly examine what information the reports include.

Look at the first page of the Home Report. Note that the chart provides labeled bars showing the child's performance. Each bar is labeled with the child's National Percentile for that skill area. When you know how to interpret them, national percentiles can be the most useful scores you encounter on reports such as this. Even when you are confronted with different tests that use different scale scores, you can always interpret percentiles the same way, regardless of the test. A percentile tells the percent of students who score at or below that level. A percentile of 25, for example, means that 25 percent of children taking the test scored at or below that score. (It also means that 75 percent of students scored above that score.) Note that the average is always at the 50th percentile.

On the right side of the graph on the first page of the report, the publisher has designated the ranges of scores that constitute average, above average, and below average. You can also use this slightly more precise key for interpreting percentiles:

PERCENTILE RANGE	LEVEL
2 and Below	Deficient
3–8	Borderline
9–23	Low Average
24–75	Average
76–97	High Average
98 and Up	Superior

The second page of the Home report provides a listing of the child's strengths and weaknesses, along with keys for mastery, partial mastery, and non-mastery of the skills. Scoring services determine these breakdowns based on the child's scores as compared with those from the national norm group.

Your child's teacher or guidance counselor will probably also receive a profile report similar to the TerraNova Individual Profile Report, shown in Figures 3 and 4. That report will be kept in your child's permanent record. The first aspect of this report to notice is that the scores are expressed both numerically and graphically.

First look at the score bands under National Percentile. Note that the scores are expressed as bands, with the actual score represented by a dot within each band. The reason we express the scores as bands is to provide an idea of the amount by which typical scores may vary for each student. That is, each band represents a

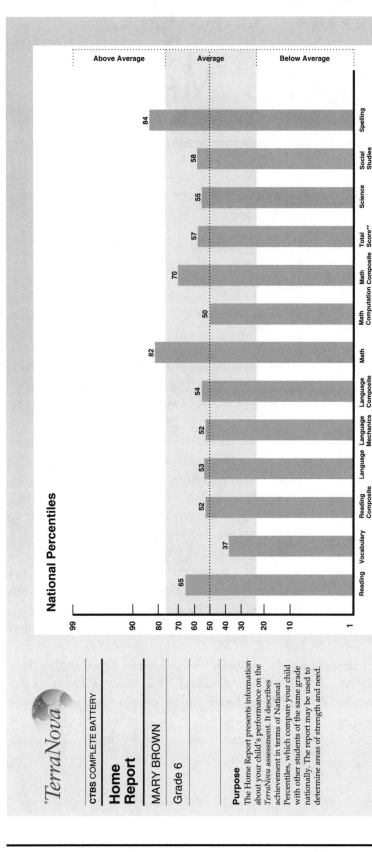

Figure 1 (SOURCE: CTB/McGraw-Hill, copyright © 1997. All rights reserved. Reproduced with permission.)

TerraNova

CTBS COMPLETE BATTERY

Home Report

MARY BROWN

Grade 6

Purpose

This page of the Home Report presents information about your child's strengths and needs. This information is provided to help you monitor your child's academic growth.

Simulated Data

Birthdate: 02/08/85
Special Codes:
A B C D E F G H I J K L M N O P Q R S T
3 5 9 7 3 2   1 1 1
Form/Level: A-16
Test Date: 11/01/99   Scoring: PATTERN (IRT)
QM: 08   Norms Date: 1996

Class: PARKER
School: WINFIELD
District: WINFIELD

City/State: WINFIELD, CA

CTB
McGraw-Hill

Page 2

Strengths

Reading
● Basic Understanding
● Analyze Text

Vocabulary
● Word Meaning
● Words in Context

Language
● Editing Skills
● Sentence Structure

Language Mechanics
● Sentences, Phrases, Clauses

Mathematics
● Computation and Numerical Estimation
● Operation Concepts

Mathematics Computation
● Add Whole Numbers
● Multiply Whole Numbers

Science
● Life Science
● Inquiry Skills

Social Studies
● Geographic Perspectives
● Economic Perspectives

Spelling
● Vowels
● Consonants

Key ● **Mastery**

General Interpretation

Needs

Reading
◐ Evaluate and Extend Meaning
○ Identify Reading Strategies

Vocabulary
○ Multimeaning Words

Language
◐ Writing Strategies

Language Mechanics
○ Writing Conventions

Mathematics
◐ Measurement
◐ Geometry and Spatial Sense

Mathematics Computation
○ Percents

Science
○ Earth and Space Science

Social Studies
◐ Historical and Cultural Perspectives

Spelling
No area of needs were identified for this content area

Key ◐ **Partial Mastery**   ○ **Non-Mastery**

The left column shows your child's best areas of performance. In each case, your child has reached mastery level. The column at the right shows the areas within each test section where your child's scores are the lowest. In these cases, your child has not reached mastery level, although he or she may have reached partial mastery.

Figure 2   (Source: CTB/McGraw-Hill, copyright © 1997. All rights reserved. Reproduced with permission.)

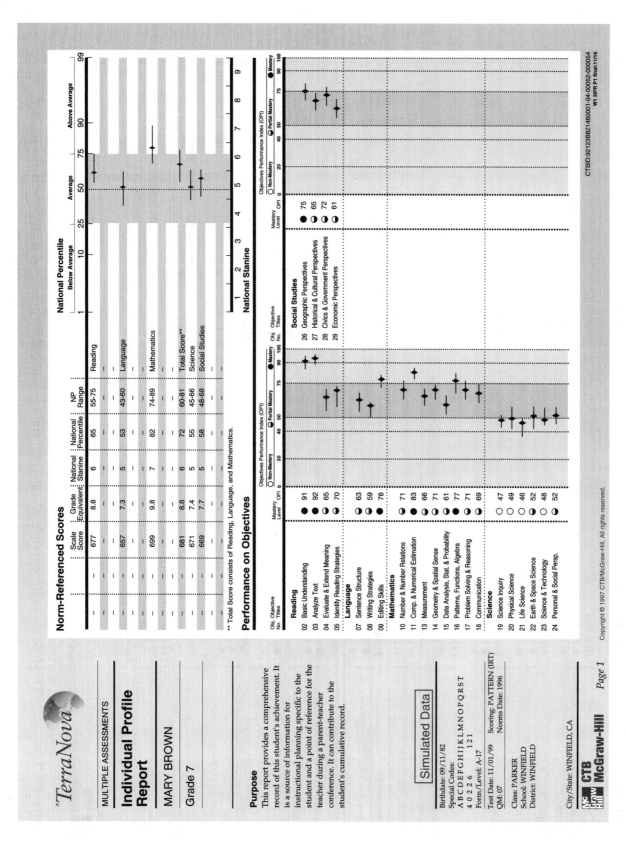

Figure 3 (SOURCE: CTB/McGraw-Hill, copyright © 1997. All rights reserved. Reproduced with permission.)

Observations

Norm-Referenced Scores

The top section of the report presents information about this student's achievement in several different ways. The National Percentile (NP) data and graph indicate how this student performed compared to students of the same grade nationally. The National Percentile range indicates that if this student had taken the test numerous times the scores would have fallen within the range shown. The shaded area on the graph represents the average range of scores, usually defined as the middle 50 percent of students nationally. Scores in the area to the right of the shading are above the average range. Scores in the area to the left of the shading are below the average range.

In Reading, for example, this student achieved a National Percentile rank of 65. This student scored higher than 65 percent of the students nationally. This score is in the average range. This student has a total of five scores in the average range. One score is in the above average range. No scores are in the below average range.

Performance on Objectives

The next section of the report presents performance on the objectives. Each objective is measured by a minimum of 4 items. The Objectives Performance Index (OPI) provides an estimate of the number of items that a student could be expected to answer correctly if there had been 100 items for that objective. The OPI is used to indicate mastery of each objective. An OPI of 75 and above characterizes Mastery. An OPI between 50 and 74 indicates Partial Mastery, and an OPI below 50 indicates Non-Mastery. The two-digit number preceding the objective title identifies the objective, which is fully described in the Teacher's Guide to *TerraNova*. The bands on either side of the diamonds indicate the range within which the student's test scores would fall if the student were tested numerous times.

In Reading, for example, this student could be expected to respond correctly to 91 out of 100 items measuring Basic Understanding. If this student had taken the test numerous times the OPI for this objective would have fallen between 82 and 93.

Teacher Notes

TerraNova

MULTIPLE ASSESSMENTS

Individual Profile Report

MARY BROWN

Grade 7

Purpose

The Observations section of the Individual Profile Report gives teachers and parents information to interpret this report. This page is a narrative description of the data on the other side.

Simulated Data

Birthdate: 09/11/82
Special Codes:
A B C D E F G H I J K L M N O P Q R S T
4 0 2 2 6   1 2 1
Form/Level: A-17

Test Date: 11/01/99   Scoring: PATTERN (IRT)
QM: 08   Norms Date: 1996

Class: PARKER
School: WINFIELD
District: WINFIELD

City/State: WINFIELD, CA

CTB
McGraw-Hill

Page 2

Figure 4 (SOURCE: CTB/McGraw-Hill, copyright © 1997. All rights reserved. Reproduced with permission.)

TerraNova

MULTIPLE ASSESSMENTS

Student Performance Level Report

KEN ALLEN

Grade 4

Purpose

This report describes this student's achievement in terms of five performance levels for each content area. The meaning of these levels is described on the back of this page. Performance levels are a new way of describing achievement.

Simulated Data

Birthdate: 02/08/86
Special Codes:
A B C D E F G H I J K L M N O P Q R S T
3 5 9 7 3 2   1 1 1
Form/Level: A-14
Test Date: 04/15/97   Scoring: PATTERN (IRT)
QM: 31   Norms Date: 1996

Class: SCHWARZ
School: WINFIELD
District: GREEN VALLEY

City/State: WINFIELD, CA

CTB McGraw-Hill   Page 1

Performance Levels	Reading	Language	Mathematics	Science	Social Studies
5 Advanced					
4 Proficient					
3 Nearing Proficiency	✓			✓	✓
2 Progressing	✓	✓	✓	✓	✓
1 Step 1	✓	✓	✓	✓	✓

Partially Proficient

Observations

Performance level scores provide a measure of what students *can do* in terms of the content and skills assessed by *TerraNova*, and typically found in curricula for Grades 3, 4, and 5. It is desirable to work towards achieving a Level 4 (Proficient) or Level 5 (Advanced) by the end of Grade 5.

The number of check marks indicates the performance level this student reached in each content area. For example, this student reached Level 3 in Reading and Social Studies.

The performance level indicates this student can perform the majority of what is described for that level and even more of what is described for the levels below. The student may also be capable of performing some of the things

described in the next higher level, but not enough to have reached that level of performance.

For example, this student can perform the majority of what is described for Level 3 in Reading and even more of what is described for Level 2 and Level 1 in Reading. This student may also be capable of performing some of what is described for Level 4 in Reading.

For each content area look at the skills and knowledge described in the next higher level. These are the competencies this student needs to demonstrate to show academic growth.

Figure 5 (Source: CTB/McGraw-Hill, copyright © 1997. All rights reserved. Reproduced with permission.)

Performance Levels (Grades 3, 4, 5)	Reading	Language	Mathematics	Science	Social Studies
5 Advanced	Students use analogies to generalize. They identify a paraphrase of concepts or ideas in texts. They can indicate thought processes that led them to a previous answer. In written responses, they demonstrate understanding of an implied theme, assess intent of passage information, and provide justification as well as support for their answers.	Students understand logical development in paragraph structure. They identify essential information from notes. They recognize the effect of prepositional phrases on subject-verb agreement. They find and correct at least 4 out of 6 errors when editing simple narratives. They correct run-on and incomplete sentences in more complex texts. They can eliminate all errors when editing their own work.	Students locate decimals on a number line; compute with decimals and fractions; read scale drawings; find areas; identify geometric transformations; construct and label bar graphs; find simple probabilities; find averages; use patterns in data to solve problems; use multiple strategies and concepts to solve unfamiliar problems; express mathematical ideas and explain the problem-solving process.	Students understand a broad range of grade level scientific concepts, such as the structure of Earth and instinctive behavior. They know terminology, such as decomposers, fossil fuel, eclipse, and buoyancy. They know the more complex environmental issues includes, for example, the positive consequences of a forest fire. Students can process and interpret more detailed tables and graphs. They can suggest improvements to experimental design, such as running more trials.	Students consistently demonstrate skills such as synthesizing information from two sources (e.g., a document and a map). They show understanding of the democratic process and global environmental issues, and know the location of continents and major countries. They analyze and summarize information from multiple sources in early American history. They thoroughly explain both sides of an issue and give complete and detailed written answers to questions.
4 Proficient	Students interpret figures of speech. They recognize paraphrase of text information and retrieve information to complete forms. In more complex texts, they identify themes, main ideas, or author purpose/point of view. They analyze and apply information in graphic and text form, make reasonable generalizations, and draw conclusions. In written responses, they can identify key elements from text.	Students select the best supporting sentences for a topic sentence. They use compound predicates to combine sentences. They identify simple subjects and predicates, recognize correct usage when confronted with two types of errors, and find and correct at least 3 out of 6 errors when editing simple narratives. They can edit their own work with only minor errors.	Students compare, order, and round whole numbers; know place value to thousands; identify fractions; use computation and estimation strategies; relate multiplication to addition; measure to nearest half-inch and centimeter; measure and find perimeters; estimate measures; find elapsed times; combine and subdivide shapes; identify parallel lines; interpret tables and graphs; solve two-step problems.	Students have a range of specific science knowledge, including details about animal adaptations and classification, states of matter, and the geology of Earth. They recognize scientific words such as habitat, gravity, and mass. They understand the usefulness of computers. They understand reasons for conserving natural resources. Understanding of experimentation includes analyzing purpose, interpreting data, and selecting tools to gather data.	Students demonstrate skills such as making inferences, using historical documents and analyzing maps to determine the economic strengths of a region. They understand the function of currency in various cultures and supply and demand. They summarize information from multiple sources, recognize relationships, determine relevance of information, and show global awareness. They propose solutions to real-world problems and support ideas with appropriate details.
3 Nearing Proficiency	Students use context clues and structural analysis to determine word meaning. They recognize homonyms and antonyms in grade-level text. They identify important details, sequence, cause and effect, and lessons embedded in the text. They interpret characters' feelings and apply information to new situations. In written responses, they can express an opinion and support it.	Students identify irrelevant sentences in paragraphs and select the best place to insert new information. They recognize faulty sentence construction. They can combine simple sentences with conjunctions and use simple subordination of phrases/clauses. They identify reference sources. They recognize correct conventions for dates, closings, and place names in informal correspondence.	Students identify even and odd numbers; subtract whole numbers with regrouping; multiply and divide by one-digit numbers; identify simple fractions; measure with ruler to nearest inch; tell time to nearest fifteen minutes; recognize and classify common shapes; recognize symmetry; subdivide shapes; complete bar graphs; extend numerical and geometric patterns; apply simple logical reasoning.	Students are familiar with the life cycles of plants and animals. They can identify an example of a cold-blooded animal. They infer what once existed from fossil evidence. They recognize the term habitat. They understand the water cycle. They know science and society issues such as recycling and sources of pollution. They can sequence technological advances. They extrapolate data, devise a simple classification scheme, and determine the purpose of a simple experiment.	Students demonstrate skills in organizing information. They use time lines, product and global maps, and cardinal directions. They understand simple cause and effect relationships and historical documents. They sequence events, associate holidays with events, and classify natural resources. They compare life in different times and understand some economic concepts related to products, jobs, and the environment. They give some detail in written responses.
2 Progressing	Students identify synonyms for grade-level words, and use context clues to define common words. They make simple inferences and predictions based on text. They identify characters' feelings. They can transfer information from text to graphic form, or from graphic form to text. In written responses, they can provide limited support for their answers.	Students identify the use of correct verb tenses and supply verbs to complete sentences. They complete paragraphs by selecting an appropriate topic sentence. They select correct adjective forms.	Students know ordinal numbers; solve coin combination problems; count by tens; add whole numbers with regrouping; have basic estimation skills; understand addition property of zero; write and identify number sentences describing simple situations; read calendars; identify appropriate measurement tools; recognize congruent figures; use simple coordinate grids; read common tables and graphs.	Students recognize that plants decompose and become part of soil. They can classify a plant as a vegetable. They recognize that camouflage relates to survival. They recognize terms such as hibernate. They have an understanding of human impact on the environment and are familiar with causes of pollution. They find the correct bar graph to represent given data and transfer data appropriate for middle elementary grades to a bar graph.	Students demonstrate simple information-processing skills such as using basic maps and keys. They recognize simple geographical terms, types of jobs, modes of transportation, and natural resources. They connect a human need with an appropriate community service. They identify some early famous presidents and know the capital of the United States. Their written answers are partially complete.
1 Step 1	Students select pictured representations of ideas and identify stated details contained in simple texts. In written responses, they can select and transfer information from charts.	Students supply subjects to complete sentences. They identify the correct use of pronouns. They edit for the correct use of end marks and initial capital letters, and identify the correct convention for greetings in letters.	Students read and recognize numbers to 1000; identify real-world use of numbers; add and subtract two-digit numbers without regrouping; identify addition situations; recognize and complete simple geometric and numerical patterns.	Students recognize basic adaptations for living in the water, identify an animal that is hatched from an egg, and associate an organism with its correct environment. They identify an object as metal. They have some understanding of conditions on the moon. They supply one way a computer can be useful. They associate an instrument like a telescope with a field of study.	Students are developing fundamental social studies skills such as locating and classifying basic information. They locate information in pictures and read and complete simple bar graphs related to social studies concepts and contexts. They can connect some city buildings with their functions and recognize certain historical objects.

Partially Proficient (levels 3, 2, 1)

W1 SPLR P2:11/02

IMPORTANT: Each performance level, depicted on the other side, indicates the student can perform the majority of what is described for that level and even more of what is described for the levels below. The student may also be capable of performing some of the things described in the next higher level, but not enough to have reached that level.

Figure 6 (SOURCE: CTB/McGraw-Hill, copyright © 1997. All rights reserved. Reproduced with permission.)

confidence interval. In these reports, we usually report either a 90 percent or 95 percent confidence interval. Interpret a confidence interval this way: Suppose we report a 90 percent confidence interval of 25 to 37. This means we estimate that, if the child took the test multiple times, we would expect that child's score to be in the 25 to 37 range 90 percent of the time.

Now look under the section titled Norm-Referenced Scores on the first page of the Individual Profile Report (Figure 3). The farthest column on the right provides the NP Range, which is the National Percentile scores represented by the score bands in the chart.

Next notice the column labeled Grade Equivalent. Theoretically, grade level equivalents equate a student's score in a skill area with the average grade placement of children who made the same score. Many psychologists and test developers would prefer that we stopped reporting grade equivalents, because they can be grossly misleading. For example, the average reading grade level of high school seniors as reported by one of the more popular tests is the eighth grade level. Does that mean that the nation's high school seniors cannot read? No. The way the test publisher calculated grade equivalents was to determine the average test scores for students in grades 4 to 6 and then simply extend the resulting prediction formula to grades 7 to 12. The result is that parents of average high school seniors who take the test in question would mistakenly believe that their seniors are reading four grade levels behind! Stick to the percentile in interpreting your child's scores.

Now look at the columns labeled Scale Score and National Stanine. These are two of a group of scores we also call *standard scores.* In reports for other tests, you may see other standard scores reported, such as Normal Curve Equivalents (NCEs), Z-Scores, and T-Scores. The IQ that we report on intelligence tests, for example, is a standard score. Standard scores are simply a way of expressing a student's scores in terms of the statistical properties of the scores from the norm group against which we are comparing the child. Although most psychologists prefer to speak in terms of standard scores among themselves, parents are advised to stick to percentiles in interpreting your child's performance.

Now look at the section of the report labeled Performance on Objectives. In this section, the test publisher reports how your child did on the various skills that make up each skills area. Note that the scores on each objective are expressed as a percentile band, and you are again told whether your child's score constitutes mastery, non-mastery, or partial mastery. Note that these scores are made up of tallies of sometimes small numbers of test items taken from sections such as Reading or Math. Because they are calculated from a much smaller number of scores than the main scales are (for example, Sentence Comprehension is made up of fewer items than overall Reading), their scores are less reliable than those of the main scales.

Now look at the second page of the Individual Profile Report (Figure 4). Here the test publisher provides a narrative summary of how the child did on the test. These summaries are computer-generated according to rules provided by the publisher. Note that the results descriptions are more general than those on the previous three report pages. But they allow the teacher to form a general picture of which students are performing at what general skill levels.

Finally, your child's guidance counselor may receive a summary report such as the TerraNova Student Performance Level Report. (See Figures 5 and 6.) In this report, the publisher explains to school personnel what skills the test assessed and generally how proficiently the child tested under each skill.

Which States Require Which Tests

Tables 1 through 3 summarize standardized testing practices in the 50 states and the District of Columbia. This information is constantly changing; the information presented here was accurate as of the date of printing of this book. Many states have changed their testing practices in response to revised accountability legislation, while others have changed the tests they use.

Table 1 State Web Sites: Education and Testing

STATE	GENERAL WEB SITE	STATE TESTING WEB SITE
Alabama	http://www.alsde.edu/	http://www.fairtest.org/states/al.htm
Alaska	www.educ.state.ak.us/	http://www.eed.state.ak.us/tls/Performance Standards/
Arizona	http://www.ade.state.az.us/	http://www.ade.state.az.us/standards/
Arkansas	http://arkedu.k12.ar.us/	http://www.fairtest.org/states/ar.htm
California	http://goldmine.cde.ca.gov/	http://ww.cde.ca.gov/cilbranch/sca/
Colorado	http://www.cde.state.co.us/index_home.htm	http://www.cde.state.co.us/index_assess.htm
Connecticut	http://www.state.ct.us/sde	http://www.state.ct.us/sde/cmt/index.htm
Delaware	http://www.doe.state.de.us/	http://www.doe.state.de.us/aab/index.htm
District of Columbia	http://www.k12.dc.us/dcps/home.html	http://www.k12.dc.us/dcps/data/data_frame2.html
Florida	http://www.firn.edu/doe/	http://www.firn.edu/doe/sas/sasshome.htm
Georgia	http://www.doe.k12.ga.us/	http://www.doe.k12.ga.us/sla/ret/recotest.html
Hawaii	http://kalama.doe.hawaii.edu/upena/	http://www.fairtest.org/states/hi.htm
Idaho	http://www.sde.state.id.us/Dept/	http://www.sde.state.id.us/instruct/ schoolaccount/statetesting.htm
Illinois	http://www.isbe.state.il.us/	http://www.isbe.state.il.us/isat/
Indiana	http://doe.state.in.us/	http://doe.state.in.us/assessment/welcome.html
Iowa	http://www.state.ia.us/educate/index.html	(Tests Chosen Locally)
Kansas	http://www.ksbe.state.ks.us/	http://www.ksbe.state.ks.us/assessment/
Kentucky	htp://www.kde.state.ky.us/	http://www.kde.state.ky.us/oaa/
Louisiana	http://www.doe.state.la.us/DOE/asps/home.asp	http://www.doe.state.la.us/DOE/asps/home.asp? I=HISTAKES
Maine	http://janus.state.me.us/education/homepage.htm	http://janus.state.me.us/education/mea/ meacompass.htm
Maryland	http://www.msde.state.md.us/	http://www.fairtest.org/states/md.htm
Massachusetts	http://www.doe.mass.edu/	http://www.doe.mass.edu/mcas/
Michigan	http://www.mde.state.mi.us/	http://www.mde.state.mi.us/off/meap/

STATE	GENERAL WEB SITE	STATE TESTING WEB SITE
Minnesota	http://www.educ.state.mn.us/	http://fairtest.org/states/mn.htm
Mississippi	http://mdek12.state.ms.us/	http://fairtest.org/states/ms.htm
Missouri	http://services.dese.state.mo.us/	http://fairtest.org/states/mo.htm
Montana	http://www.metnet.mt.gov/	http://fairtest.org/states/mt.htm
Nebraska	http://nde4.nde.state.ne.us/	http://www.edneb.org/IPS/AppAccrd/ApprAccrd.html
Nevada	http://www.nsn.k12.nv.us/nvdoe/	http://www.nsn.k12.nv.us/nvdoe/reports/TerraNova.doc
New Hampshire	http://www.state.nh.us/doe/	http://www.state.nh.us/doe/Assessment/assessme(NHEIAP).htm
New Jersey	http://ww.state.nj.us/education/	http://www.state.nj.us/njded/stass/index.html
New Mexico	http://sde.state.nm.us/	http://sde.state.nm.us/press/august30a.html
New York	http://www.nysed.gov/	http://www.emsc.nysed.gov/ciai/assess.html
North Carolina	http://www.dpi.state.nc.us/	http://www.dpi.state.nc.us/accountability/reporting/index.html
North Dakota	http://www.dpi.state.nd.us/dpi/index.htm	http://www.dpi.state.nd.us/dpi/reports/assess/assess.htm
Ohio	http://www.ode.state.oh.us/	http://www.ode.state.oh.us/ca/
Oklahoma	http://sde.state.ok.us/	http://sde.state.ok.us/acrob/testpack.pdf
Oregon	http://www.ode.state.or.us//	http://www.ode.state.or.us/assmt/index.htm
Pennsylvania	http://www.pde.psu.edu/ http://instruct.ride.ri.net/ride_home_page.html	http://www.fairtest.org/states/pa.htm
Rhode Island		
South Carolina	http://www.state.sc.us/sde/	http://www.state.sc.us/sde/reports/terranov.htm
South Dakota	http://www.state.sd.us/state/executive/deca/	http://www.state.sd.us/state/executive/deca/TA/McRelReport/McRelReports.htm
Tennessee	http://www.state.tn.us/education/	http://www.state.tn.us/education/tsintro.htm
Texas	http://www.tea.state.tx.us/	http://www.tea.state.tx.us/student.assessment/
Utah	http://www.usoe.k12.ut.us/	http://www.usoe.k12.ut.us/eval.usoeeval.htm
Vermont	http://www.cit.state.vt.us/educ/	http://www.fairtest.org/states/vt.htm

STATE	GENERAL WEB SITE	STATE TESTING WEB SITE
Virginia	http://www.pen.k12.va.us/Anthology/VDOE/	http://www.pen.k12.va.us/VDOE/Assessment/home.shtml
Washington	http://www.k12.wa.us/	http://assessment.ospi.wednet.edu/
West Virginia	http://wvde.state.wv.us/	http://www.fairtest.org/states/wv.htm
Wisconsin	http://www.dpi.state.wi.us/	http://www.dpi.state.wi.us/dpi/oea/spr_kce.html
Wyoming	http://www.k12.wy.us/wdehome.html	http://www.asme.com/wycas/index.htm

Table 2 Norm-Referenced and Criterion-Referenced Tests Administered by State

STATE	NORM-REFERENCED TEST	CRITERION-REFERENCED TEST	EXIT EXAM
Alabama	Stanford Achievement Test		Alabama High School Graduation Exam
Alaska	California Achievement Test		
Arizona	Stanford Achievement Test	Arizona's Instrument to Measure Standards (AIMS)	
Arkansas	Stanford Achievement Test		
California	Stanford Achievement Test	Standardized Testing and Reporting Supplement	High School Exit Exam (HSEE)
Colorado	None	Colorado Student Assessment Program	
Connecticut		Connecticut Mastery Test	
Delaware	Stanford Achievement Test	Delaware Student Testing Program	
District of Columbia	Stanford Achievement Test		
Florida	(Locally Selected)	Florida Comprehensive Assessment Test (FCAT)	High School Competency Test (HSCT)
Georgia	Iowa Tests of Basic Skills	Criterion-Referenced Competency Tests (CRCT)	Georgia High School Graduation Tests
Hawaii	Stanford Achievement Test	Credit by Examination	Hawaii State Test of Essential Competencies
Idaho	Iowa Test of Basic Skills/ Tests of Direct Achievement and Proficiency	Writing/Mathematics Assessment	
Illinois		Illinois Standards Achievement Tests	Prairie State Achievement Examination
Indiana		Indiana Statewide Testing for Education Progress	
Iowa	(None)		
Kansas		(State-Developed Tests)	
Kentucky	Comprehensive Tests of Basic Skills	Kentucky Instructional Results Information System	
Louisiana	Iowa Tests of Basic Skills	Louisiana Educational Assessment Program	Graduate Exit Exam
Maine		Maine Educational Assessment	
Maryland		Maryland School Performance Assessment Program	
Massachusetts		Massachusetts Comprehensive Assessment System	

STATE	NORM-REFERENCED TEST	CRITERION-REFERENCED TEST	EXIT EXAM
Michigan		Michigan Educational Assessment Program	High School Test
Minnesota		Basic Standards Test	Profile of Learning
Mississippi	Iowa Test of Basic Skills	Subject Area Testing Program	Functional Literacy Examination
Missouri		Missouri Mastery and Achievement Test	
Montana	(districts' choice)		
Nebraska			
Nevada	TerraNova		Nevada High School Proficiency Examination
New Hampshire		NH Educational Improvement and Assessment Program	
New Jersey		Elementary School Proficiency Test/Early Warning Test	High School Proficiency Test
New Mexico	TerraNova		New Mexico High School Competency Exam
New York		Pupil Evaluation Program/ Preliminary Competency Test	Regents Competency Tests
North Carolina	Iowa Test of Basic Skills	NC End of Grade Test	
North Dakota	TerraNova	ND Reading, Writing Speaking, Listening, Math Test	
Ohio		Ohio Proficiency Tests	Ohio Proficiency Tests
Oklahoma	Iowa Tests of Basic Skills	Oklahoma Criterion-Referenced Tests	
Oregon		Oregon Statewide Assessment	
Pennsylvania		Pennsylvania System of School Assessment	
Rhode Island	Metropolitan Achievement Test		
South Carolina	TerraNova	Palmetto Achievement Challenge Tests	High School Exit Exam
South Dakota	Stanford Achievement Test		
Tennessee	Tennessee Comprehensive Assessment Program	Tennessee Comprehensive Assessment Program	
Texas		Texas Assessment of Academic Skills	Texas Assessment of Academic Skills
Utah	Stanford Achievement Test	Core Curriculum Testing	

STATE	NORM-REFERENCED TEST	CRITERION-REFERENCED TEST	EXIT EXAM
Vermont		New Standards Reference Exams	
Virginia	Stanford Achievement Test	Virginia Standards of Learning	Virginia Standards of Learning
Washington	Iowa Tests of Basic Skills	Washington Assessment of Student Learning	Washington Assessment of Student Learning
West Virginia	Stanford Achievement Test		
Wisconsin	TerraNova	Wisconsin Knowledge and Concepts Examinations	
Wyoming	TerraNova	Wyoming Comprehensive Assessment System	Wyoming Comprehensive Assessment System

Table 3 Standardized Test Schedules by State

STATE	KG	1	2	3	4	5	6	7	8	9	10	11	12	COMMENT
Alabama				X	X	X	X	X	X	X	X	X	X	
Alaska					X				X			X		
Arizona			X	X	X	X	X	X	X	X	X	X	X	
Arkansas					X	X		X	X		X	X	X	
California			X	X	X	X	X	X	X	X	X	X		
Colorado				X	X			X						
Connecticut					X		X		X					
Delaware				X	X	X			X		X	X		
District of Columbia		X	X	X	X	X	X	X	X	X	X	X		
Florida		X	X	X	X	X	X	X	X	X	X	X	X	There is no state-mandated norm-referenced testing. However, the state collects information furnished by local districts that elect to perform norm-referenced testing. The FCAT is administered to Grades 4, 8, and 10 to assess reading and Grades 5, 8, and 10 to assess math.
Georgia				X		X			X					
Hawaii				X			X		X		X			The Credit by Examination is voluntary and is given in Grade 8 in Algebra and Foreign Languages.
Idaho				X	X	X	X	X	X	X	X	X		
Illinois				X	X		X	X	X		X	X		Exit Exam failure will not disqualify students from graduation if all other requirements are met.
Indiana				X			X		X		X			
Iowa		*	*	*	*	*	*	*	*	*	*	*	*	*Iowa does not currently have a statewide testing program. Locally chosen assessments are administered to grades determined locally.
Kansas				X	X	X		X	X		X			

STATE	KG	1	2	3	4	5	6	7	8	9	10	11	12	COMMENT
Kentucky					X	X		X	X			X	X	
Louisiana				X		X	X	X		X				
Maine					X				X			X		
Maryland				X		X			X					
Massachusetts					X				X		X			
Michigan					X	X		X	X					
Minnesota				X		X			X					Testing Information from Fair Test.Org. There was no readily accessible state-sponsored site.
Mississippi					X	X	X	X	X					State's Web site refused connection; all data were obtained from FairTest.Org.
Missouri			X	X	X	X	X	X	X	X	X			
Montana					X				X			X		The State Board of Education has decided to use a single norm-referenced test statewide beginning 2000–2001 school year.
Nebraska		**	**	**	**	**	**	**	**	**	**	**	**	**Decisions regarding testing are left to the individual school districts.
Nevada					X				X					Districts choose whether and how to test with norm-referenced tests.
New Hampshire				X			X				X			
New Jersey				X	X			X	X	X	X	X		
New Mexico					X		X		X					
New York					X				X	X				Assessment program is going through major revisions.
North Carolina				X	X	X	X	X	X		X			NRT Testing selects samples of students, not all.
North Dakota					X		X		X		X			
Ohio					X		X			X			X	
Oklahoma				X		X		X	X			X		
Oregon				X		X			X		X			

STATE	KG	1	2	3	4	5	6	7	8	9	10	11	12	COMMENT
Pennsylvania						X	X		X	X		X		
Rhode Island				X	X	X		X	X	X	X			
South Carolina				X	X	X	X	X	X	X	X			
South Dakota			X		X	X			X	X		X		
Tennessee			X	X	X	X	X	X	X			X		
Texas				X	X	X	X	X	X		X			
Utah		X	X	X	X	X	X	X	X	X	X	X	X	
Vermont					X	X	X		X	X	X	X		Rated by FairTest.Org as a nearly model system for assessment.
Virginia				X	X	X	X		X	X		X		
Washington					X			X			X			
West Virginia		X	X	X	X	X	X	X	X	X	X	X		
Wisconsin					X				X		X			
Wyoming					X				X			X		

Testing Accommodations

The more testing procedures vary from one classroom or school to the next, the less we can compare the scores from one group to another. Consider a test in which the publisher recommends that three sections of the test be given in one 45-minute session per day on three consecutive days. School A follows those directions. To save time, School B gives all three sections of the test in one session lasting slightly more than two hours. We can't say that both schools followed the same testing procedures. Remember that the test publishers provide testing procedures so schools can administer the tests in as close a manner as possible to the way the tests were administered to the groups used to obtain test norms. When we compare students' scores to norms, we want to compare apples to apples, not apples to oranges.

Most schools justifiably resist making any changes in testing procedures. Informally, a teacher can make minor changes that don't alter the testing procedures, such as separating two students who talk with each other instead of paying attention to the test; letting Lisa, who is getting over an ear infection, sit closer to the front so she can hear better; or moving Jeffrey away from the window to prevent his looking out the window and daydreaming.

There are two groups of students who require more formal testing accommodations. One group of students is identified as having a disability under Section 504 of the Rehabilitation Act of 1973 (Public Law 93-112). These students face some challenge but, with reasonable and appropriate accommodation, can take advantage of the same educational opportunities as other students. That is, they have a condition that requires some accommodation for them.

Just as schools must remove physical barriers to accommodate students with disabilities, they must make appropriate accommodations to remove other types of barriers to students' access to education. Marie is profoundly deaf, even with strong hearing aids. She does well in school with the aid of an interpreter, who signs her teacher's instructions to her and tells her teacher what Marie says in reply. An appropriate accommodation for Marie would be to provide the interpreter to sign test instructions to her, or to allow her to watch a videotape with an interpreter signing test instructions. Such a reasonable accommodation would not deviate from standard testing procedures and, in fact, would ensure that Marie received the same instructions as the other students.

If your child is considered disabled and has what is generally called a Section 504 Plan or individual accommodation plan (IAP), then the appropriate way to ask for testing accommodations is to ask for them in a meeting to discuss school accommodations under the plan. If your child is not already covered by such a plan, he or she won't qualify for one merely because you request testing accommodations.

The other group of students who may receive formal testing accommodations are those identified as handicapped under the Individuals with Disabilities Education Act (IDEA)—students with mental retardation, learning disabilities, serious emotional disturbance, orthopedic handicap, hearing or visual problems, and other handicaps defined in the law. These students have been identified under procedures governed by federal and sometimes state law, and their education is governed by a document called the Individualized Educational Program (IEP). Unless you are under a court order specifically revoking your educational rights on behalf of your child, you are a full member of the IEP team even if you and your child's other parent are divorced and the other parent has custody. Until recently, IEP teams actually had the prerogative to exclude certain handicapped students from taking standardized group testing altogether. However, today states make it more difficult to exclude students from testing.

If your child is classified as handicapped and has an IEP, the appropriate place to ask for testing accommodations is in an IEP team meeting. In fact, federal regulations require IEP teams to address testing accommodations. You have the right to call a meeting at any time. In that meeting, you will have the opportunity to present your case for the accommodations you believe are necessary. Be prepared for the other team members to resist making extreme accommodations unless you can present a very strong case. If your child is identified as handicapped and you believe that he or she should be provided special testing accommodations, contact the person at your child's school who is responsible for convening IEP meetings and request a meeting to discuss testing accommodations.

Problems arise when a request is made for accommodations that cause major departures from standard testing procedures. For example, Lynn has an identified learning disability in mathematics calculation and attends resource classes for math. Her disability is so severe that her IEP calls for her to use a calculator when performing all math problems. She fully understands math concepts, but she simply can't perform the calculations without the aid of a calculator. Now it's time for Lynn to take the school-based standardized tests, and she asks to use a calculator. In this case, since her IEP already requires her to be provided with a calculator when performing math calculations, she may be allowed a calculator during school standardized tests. However, because using a calculator constitutes a major violation of standard testing procedures, her score on all sections in which she is allowed to use a calculator will be recorded as a failure, and her results in some states will be removed from among those of other students in her school in calculating school results.

How do we determine whether a student is allowed formal accommodations in standardized school testing and what these accommodations may be? First, if your child is not already identified as either handicapped or disabled, having the child classified in either group solely to receive testing accommodations will be considered a violation of the laws governing both classifications. Second, even if your child is already classified in either group, your state's department of public instruction will provide strict guidelines for the testing accommodations schools may make. Third, even if your child is classified in either group and you are proposing testing accommodations allowed under state testing guidelines, any accommodations must still be both *reasonable* and *appropriate*. To be reasonable and appropriate, testing accommodations must relate to your child's disability and must be similar to those already in place in his or her daily educational program. If your child is always tested individually in a separate room for all tests in all subjects, then a similar practice in taking school-based standardized tests may be appropriate. But if your child has a learning disability only in mathematics calculation, requesting that all test questions be read to him or her is inappropriate because that accommodation does not relate to his identified handicap.

Glossary

Accountability The idea that a school district is held responsible for the achievement of its students. The term may also be applied to holding students responsible for a certain level of achievement in order to be promoted or to graduate.

Achievement test An assessment that measures current knowledge in one or more of the areas taught in most schools, such as reading, math, and language arts.

Aptitude test An assessment designed to predict a student's potential for learning knowledge or skills.

Content validity The extent to which a test represents the content it is designed to cover.

Criterion-referenced test A test that rates how thoroughly a student has mastered a specific skill or area of knowledge. Typically, a criterion-referenced test is subjective, and relies on someone to observe and rate student work; it doesn't allow for easy comparisons of achievement among students. Performance assessments are criterion-referenced tests. The opposite of a criterion-referenced test is a norm-referenced test.

Frequency distribution A tabulation of individual scores (or groups of scores) that shows the number of persons who obtained each score.

Generalizability The idea that the score on a test reflects what a child knows about a subject, or how well he performs the skills the test is supposed to be assessing. Generalizability requires that enough test items are administered to truly assess a student's achievement.

Grade equivalent A score on a scale developed to indicate the school grade (usually measured in months of a year) that corresponds to an average chronological age, mental age, test score, or other characteristic. A grade equivalent of 6.4 is interpreted as a score that is average for a group in the fourth month of Grade 6.

High-stakes assessment A type of standardized test that has major consequences for a student or school (such as whether a child graduates from high school or gets admitted to college).

Mean Average score of a group of scores.

Median The middle score in a set of scores ranked from smallest to largest.

National percentile Percentile score derived from the performance of a group of individuals across the nation.

Normative sample A comparison group consisting of individuals who have taken a test under standard conditions.

Norm-referenced test A standardized test that can compare scores of students in one school with a reference group (usually other students in the same grade and age, called the "norm group"). Norm-referenced tests compare the achievement of one student or the students of a school, school district, or state with the norm score.

Norms A summary of the performance of a group of individuals on which a test was standardized.

Percentile An incorrect form of the word *centile,* which is the percent of a group of scores that falls below a given score. Although the correct term is *centile,* much of the testing literature has adopted the term *percentile.*

Performance standards A level of performance on a test set by education experts.

Quartiles Points that divide the frequency distribution of scores into equal fourths.

Regression to the mean The tendency of scores in a group of scores to vary in the direction of the mean. For example: If a child has an abnormally low score on a test, she is likely to make a higher score (that is, one closer to the mean) the next time she takes the test.

Reliability The consistency with which a test measures some trait or characteristic. A measure can be reliable without being valid, but it can't be valid without being reliable.

Standard deviation A statistical measure used to describe the extent to which scores vary in a group of scores. Approximately 68 percent of scores in a group are expected to be in a range from one standard deviation below the mean to one standard deviation above the mean.

Standardized test A test that contains well-defined questions of proven validity and that produces reliable scores. Such tests are commonly paper-and-pencil exams containing multiple-choice items, true or false questions, matching exercises, or short fill-in-the-blanks items. These tests may also include performance assessment items (such as a writing sample), but assessment items cannot be completed quickly or scored reliably.

Test anxiety Anxiety that occurs in test-taking situations. Test anxiety can seriously impair individuals' ability to obtain accurate scores on a test.

Validity The extent to which a test measures the trait or characteristic it is designed to measure. Also see *reliability.*

Answer Keys for Practice Skills

Chapter 2:
Vocabulary

1	B
2	C
3	A
4	D
5	B
6	C
7	D
8	A
9	B
10	C
11	A
12	D
13	C
14	A
15	C
16	B
17	C
18	A
19	C
20	C
21	B
22	C
23	D

Chapter 3:
Reading
Comprehension

1	C
2	B
3	B
4	A
5	D
6	B
7	A
8	D
9	C
10	D
11	A
12	B
13	B
14	C
15	A
16	D

Chapter 4:
Language Mechanics

1	B
2	C
3	B
4	A
5	A
6	C
7	C
8	C
9	B
10	C
11	A
12	B
13	C
14	A
15	B
16	D
17	C
18	A
19	C
20	C
21	B
22	D
23	A
24	B
25	C

Chapter 5:
Language Expression

1	B
2	B
3	C
4	A
5	B
6	B
7	B
8	C
9	D
10	B
11	A
12	A
13	B
14	A
15	B
16	B
17	C
18	C
19	A
20	C

21	B
22	B
23	B
24	A
25	D
26	B
27	A
28	B
29	A

Chapter 6:
Spelling and Study Skills

1	B
2	A
3	C
4	A
5	D
6	C
7	B
8	A
9	B
10	B
11	D
12	B
13	C
14	C
15	A
16	B
17	B
18	D
19	B

20	A
21	C
22	B
23	C
24	C

Chapter 7:
Math Concepts

1	B
2	C
3	B
4	B
5	D
6	D
7	A
8	C
9	B
10	C
11	C

12	B
13	B
14	D
15	A
16	C
17	B
18	B
19	D
20	C
21	C
22	C

Chapter 8:
Math Computation

1	B
2	D
3	A
4	B
5	C

6	B
7	D
8	E
9	A
10	C
11	C
12	B
13	C
14	C
15	C
16	C
17	B

Chapter 9:
Math Applications

1	B
2	C
3	C
4	D

5	A
6	C
7	B
8	B
9	C
10	B
11	D
12	D
13	A
14	C
15	B

Sample Practice Test

The test questions given here are designed to provide a sample of the kinds of items sixth-grade students may encounter on a standardized test. They are *not* identical to any standardized test your child will take. However, the questions cover all the areas discussed in this book and provide a review that is similar in format to a standardized test.

The sample test provides 109 questions organized by skill areas presented in the preceding chapters. It is intended to provide a rough idea about the types of test questions your child will probably encounter on the commercial standardized tests provided at school. It is not an exact copy.

How to Use the Test

In this guide, we have been more concerned with strengthening certain skills than with the ability to work under time constraints. We don't recommend that you attempt to simulate actual testing conditions. Here are four alternative ways of using this test:

1. Administer these tests to your child after you have completed all skills chapters and have begun to implement the strategies we suggested. Allow your child to work at a leisurely pace, probably consisting of 20- to 30-minute sessions spread out over several days.

2. Administer the pertinent section of the test after you have been through each chapter and implemented the strategies.

3. Use the tests as a pretest rather than as a posttest, administering the entire test in 20- to 30-minute sessions spread out over several days to identify the skills on which your child needs the most work. Then concentrate most of your efforts on the skills on which your child scores the lowest.

4. Administer each section of the test before you go through each chapter as a kind of skills check to help you determine how much of your energy you need to devote to that skill area.

Administering the Test

Don't provide any help to your child during these tests, but note specific problems. For example, if your child has problems reading math sentences, note whether the problem is with reading rather than with math. If your child's answers look sloppy, with many erasures or cross-outs, note that you need to work on neatness. (Remember that tests administered at school will be machine-scored, and the scanners sometimes mistake sloppily erased answers as the answers the child intends.)

To the Student:

These tests will give you a chance to put the tips you have learned to work.

A few last reminders . . .

- Be sure you understand all the directions before you begin each test. You may ask the teacher questions about the directions if you do not understand them.
- Work as quickly as you can during each test.
- When you change an answer, be sure to erase your first mark completely.

- You can guess at an answer or skip difficult items and go back to them later.
- Use the tips you have learned whenever you can.
- It is OK to be a little nervous. You may even do better.

Now that you have completed the lessons in this book, you are on your way to scoring high!

STUDENT'S NAME		SCHOOL
LAST	FIRST	MI	TEACHER

FEMALE ◯   MALE ◯

BIRTHDATE

MONTH	DAY	YEAR

JAN ◯   FEB ◯   MAR ◯   APR ◯   MAY ◯   JUN ◯   JUL ◯   AUG ◯   SEP ◯   OCT ◯   NOV ◯   DEC ◯

DAY: ⓪ ⓪ / ① ① / ② ② / ③ ③ / ④ / ⑤ / ⑥ / ⑦ / ⑧ / ⑨

YEAR: ⓪ / ① / ② / ③ / ④ / ⑤ ⑤ / ⑥ ⑥ / ⑦ ⑦ / ⑧ ⑧ / ⑨ ⑨

GRADE: ① ② ③ ④ ⑤ ⑥

(Name fields: columns of bubbles A B C D E F G H I J K L M N O P Q R S T U V W X Y Z)

Vocabulary

1 Ⓐ Ⓑ Ⓒ Ⓓ	5 Ⓐ Ⓑ Ⓒ Ⓓ	9 Ⓐ Ⓑ Ⓒ Ⓓ	13 Ⓐ Ⓑ Ⓒ Ⓓ	16 Ⓐ Ⓑ Ⓒ Ⓓ
2 Ⓐ Ⓑ Ⓒ Ⓓ	6 Ⓐ Ⓑ Ⓒ Ⓓ	10 Ⓐ Ⓑ Ⓒ Ⓓ	14 Ⓐ Ⓑ Ⓒ Ⓓ	17 Ⓐ Ⓑ Ⓒ Ⓓ
3 Ⓐ Ⓑ Ⓒ Ⓓ	7 Ⓐ Ⓑ Ⓒ Ⓓ	11 Ⓐ Ⓑ Ⓒ Ⓓ	15 Ⓐ Ⓑ Ⓒ Ⓓ	18 Ⓐ Ⓑ Ⓒ Ⓓ
4 Ⓐ Ⓑ Ⓒ Ⓓ	8 Ⓐ Ⓑ Ⓒ Ⓓ	12 Ⓐ Ⓑ Ⓒ Ⓓ		

Reading Comprehension

1 Ⓐ Ⓑ Ⓒ Ⓓ	5 Ⓐ Ⓑ Ⓒ Ⓓ	9 Ⓐ Ⓑ Ⓒ Ⓓ	12 Ⓐ Ⓑ Ⓒ Ⓓ	15 Ⓐ Ⓑ Ⓒ Ⓓ
2 Ⓐ Ⓑ Ⓒ Ⓓ	6 Ⓐ Ⓑ Ⓒ Ⓓ	10 Ⓐ Ⓑ Ⓒ Ⓓ	13 Ⓐ Ⓑ Ⓒ Ⓓ	16 Ⓐ Ⓑ Ⓒ Ⓓ
3 Ⓐ Ⓑ Ⓒ Ⓓ	7 Ⓐ Ⓑ Ⓒ Ⓓ	11 Ⓐ Ⓑ Ⓒ Ⓓ	14 Ⓐ Ⓑ Ⓒ Ⓓ	17 Ⓐ Ⓑ Ⓒ Ⓓ
4 Ⓐ Ⓑ Ⓒ Ⓓ	8 Ⓐ Ⓑ Ⓒ Ⓓ			

Language Mechanics

1 Ⓐ Ⓑ Ⓒ Ⓓ	4 Ⓐ Ⓑ Ⓒ Ⓓ	7 Ⓐ Ⓑ Ⓒ Ⓓ	9 Ⓐ Ⓑ Ⓒ Ⓓ	11 Ⓐ Ⓑ Ⓒ Ⓓ
2 Ⓐ Ⓑ Ⓒ Ⓓ	5 Ⓐ Ⓑ Ⓒ Ⓓ	8 Ⓐ Ⓑ Ⓒ Ⓓ	10 Ⓐ Ⓑ Ⓒ Ⓓ	12 Ⓐ Ⓑ Ⓒ Ⓓ
3 Ⓐ Ⓑ Ⓒ Ⓓ	6 Ⓐ Ⓑ Ⓒ Ⓓ			

Language Expression

1 Ⓐ Ⓑ Ⓒ Ⓓ	4 Ⓐ Ⓑ Ⓒ Ⓓ	7 Ⓐ Ⓑ Ⓒ Ⓓ	10 Ⓐ Ⓑ Ⓒ Ⓓ	12 Ⓐ Ⓑ Ⓒ Ⓓ
2 Ⓐ Ⓑ Ⓒ Ⓓ	5 Ⓐ Ⓑ Ⓒ Ⓓ	8 Ⓐ Ⓑ Ⓒ Ⓓ	11 Ⓐ Ⓑ Ⓒ Ⓓ	13 Ⓐ Ⓑ Ⓒ Ⓓ
3 Ⓐ Ⓑ Ⓒ Ⓓ	6 Ⓐ Ⓑ Ⓒ Ⓓ	9 Ⓐ Ⓑ Ⓒ Ⓓ		

Spelling and Study Skills

1 Ⓐ Ⓑ Ⓒ Ⓓ	4 Ⓐ Ⓑ Ⓒ Ⓓ	7 Ⓐ Ⓑ Ⓒ Ⓓ	9 Ⓐ Ⓑ Ⓒ Ⓓ	11 Ⓐ Ⓑ Ⓒ Ⓓ
2 Ⓐ Ⓑ Ⓒ Ⓓ	5 Ⓐ Ⓑ Ⓒ Ⓓ	8 Ⓐ Ⓑ Ⓒ Ⓓ	10 Ⓐ Ⓑ Ⓒ Ⓓ	12 Ⓐ Ⓑ Ⓒ Ⓓ
3 Ⓐ Ⓑ Ⓒ Ⓓ	6 Ⓐ Ⓑ Ⓒ Ⓓ			

Math Concepts

1 Ⓐ Ⓑ Ⓒ Ⓓ	4 Ⓐ Ⓑ Ⓒ Ⓓ	7 Ⓐ Ⓑ Ⓒ Ⓓ	9 Ⓐ Ⓑ Ⓒ Ⓓ	11 Ⓐ Ⓑ Ⓒ Ⓓ
2 Ⓐ Ⓑ Ⓒ Ⓓ	5 Ⓐ Ⓑ Ⓒ Ⓓ	8 Ⓐ Ⓑ Ⓒ Ⓓ	10 Ⓐ Ⓑ Ⓒ Ⓓ	12 Ⓐ Ⓑ Ⓒ Ⓓ
3 Ⓐ Ⓑ Ⓒ Ⓓ	6 Ⓐ Ⓑ Ⓒ Ⓓ			

Math Computation

1 Ⓐ Ⓑ Ⓒ Ⓓ	4 Ⓐ Ⓑ Ⓒ Ⓓ	7 Ⓐ Ⓑ Ⓒ Ⓓ	10 Ⓐ Ⓑ Ⓒ Ⓓ	12 Ⓐ Ⓑ Ⓒ Ⓓ
2 Ⓐ Ⓑ Ⓒ Ⓓ	5 Ⓐ Ⓑ Ⓒ Ⓓ	8 Ⓐ Ⓑ Ⓒ Ⓓ	11 Ⓐ Ⓑ Ⓒ Ⓓ	13 Ⓐ Ⓑ Ⓒ Ⓓ
3 Ⓐ Ⓑ Ⓒ Ⓓ	6 Ⓐ Ⓑ Ⓒ Ⓓ	9 Ⓐ Ⓑ Ⓒ Ⓓ		

Math Applications

1 Ⓐ Ⓑ Ⓒ Ⓓ	4 Ⓐ Ⓑ Ⓒ Ⓓ	7 Ⓐ Ⓑ Ⓒ Ⓓ	9 Ⓐ Ⓑ Ⓒ Ⓓ	11 Ⓐ Ⓑ Ⓒ Ⓓ
2 Ⓐ Ⓑ Ⓒ Ⓓ	5 Ⓐ Ⓑ Ⓒ Ⓓ	8 Ⓐ Ⓑ Ⓒ Ⓓ	10 Ⓐ Ⓑ Ⓒ Ⓓ	12 Ⓐ Ⓑ Ⓒ Ⓓ
3 Ⓐ Ⓑ Ⓒ Ⓓ	6 Ⓐ Ⓑ Ⓒ Ⓓ			

VOCABULARY

Directions: Read each item. Choose the word that means the same or nearly the same as the underlined word.

Example:

The competition was quite <u>fierce.</u>

A   angry
B   tough
C   easy
D   busy

Answer:

B   tough

---

1   ascend the <u>knoll</u>

A   mountain
B   hill
C   range
D   lake

---

2   <u>duly</u> constituted

A   two
B   lackluster
C   properly
D   gloomily

---

3   <u>galled</u> by the suggestion

A   irritated
B   energized
C   groomed
D   perplexed

Directions: Read each item. Choose the word that means the opposite of the underlined word.

Example:

<u>pierced</u> the armor

A   tickled
B   perforated
C   glanced off
D   hit

Answer:

C   glanced off

---

4   <u>accelerated</u> the car

A   wrecked
B   accented
C   purchased
D   braked

---

5   the <u>concluding</u> chapter

A   final
B   beginning
C   middle
D   important

---

6   an <u>ornery</u> man

A   cross
B   grouchy
C   agreeable
D   simple

Directions: For questions 7 through 9, read the directions carefully and choose the answer you believe is correct.

GO

Example:

The dog loved to <u>run</u> through the fields.

In which sentence does the word <u>run</u> mean the same thing as in the sentence above?

A   Sally had a <u>run</u> in her stocking.

B   We wanted to <u>run</u> down the hill.

C   We went fishing in the mill <u>run.</u>

D   I had the <u>run</u> of the shop while the boss was gone.

Answer:

B   We wanted to <u>run</u> down the hill.

---

7   She was about to <u>type</u> the report on her computer.

In which sentence does the word <u>type</u> mean the same as in the sentence above?

A   She had to check her blood <u>type</u> before she could get the transfusion.

B   She had to know what <u>type</u> of blouse to wear to the concert.

C   He likes to <u>type</u> his essays as he writes them.

D   Let's <u>type</u> these viruses before we continue with the experiment so that we can identify them.

---

8   During the <u>break,</u> we had time to visit and buy a soft drink.

In which sentence does the word <u>break</u> mean the same as in the sentence above?

A   We found a <u>break</u> in the water line leading to our house.

B   She took a <u>break</u> from studying and called her best friend on the phone.

C   After the examination, the doctor pointed out the place of the <u>break</u> in her foot.

D   She did not want to <u>break</u> anything, so she worked with the dishes very carefully.

---

9   I hope I have a <u>chance</u> to see her again soon.

In which sentence does the word <u>chance</u> mean the same as in the sentence above?

A   We took a <u>chance</u> on the raffle, hoping we would win.

B   She enjoyed playing games of <u>chance</u> at the party.

C   What is the <u>chance</u> that you will be in the area during the holiday?

D   We enjoyed the <u>chance</u> to eat at that restaurant again so soon.

Directions: Read the paragraph. Find the word below the paragraph that fits in each numbered blank.

The following passage is taken from *The Call of the Wild* by Jack London:

His father, Elmo, a huge St. Bernard, had been the Judge's inseparable companion, and Buck bid fair to follow in the way of his father. He was not so large—he weighed only one hundred and forty pounds—for his mother, Shep, had been a Scotch shepherd dog. Nevertheless, one hundred and forty pounds, to which was added the ___10___ that comes of good living and universal respect, enabled him to carry himself in right royal fashion. During the four years since his puppyhood he had lived the life of a sated ___11___; he had a fine pride in himself, was even a trifle ___12___, as country gentlemen sometimes become because of their insular situation. But he had saved himself by not becoming a mere pampered house dog. Hunting and kindred outdoor delights had kept down the fat and hardened his muscles; and to him, as to the cold-tubbing races, the love of water had been a tonic and a health ___13___.

10 **A** hunger
    B dignity
    C size
    D food

11 **A** pet
    B killer
    C aristocrat
    D servant

12 **A** egotistical
    B humble
    C small
    D frantic

13 **A** risk
    B concern
    C exercise
    D preserver

Directions: Choose the correct answer for questions 14 and 15.

Example:

Which of the following comes from the Latin word <u>gener</u> meaning birth, race, kind?

    A generous
    B genus
    C giant
    D geode

Answer:

    B genus

14 Choose the answer that best defines the underlined part.

    aquar<u>ium</u>   planetar<u>ium</u>

    A person who
    B place where
    C study of
    D love of

15 Which of the following words comes from the Middle French word <u>rehercier</u> meaning "to repeat"?

    A rehearse
    B regulate
    C register
    D rehabilitate

Directions: Choose the answer that means the same as the underlined word.

Example:

the potato was <u>gigantic</u>

    A incredibly green
    B extremely large
    C very ripe
    D delicious

Answer:

    B extremely large

16 an overnight <u>vigil</u>

    A a period of sickness
    B a time of great evil
    C a period of watchfulness
    D a time of sleeping

17 <u>vehemently</u> denying

    A showing great force or feeling
    B without conviction
    C slowly
    D showing disinterest

18 <u>perpetual</u> motion

    A slow
    B fast
    C short-lived
    D lasting forever

STOP

READING COMPREHENSION

Directions: Read this passage from *Little Women* by Louisa May Alcott and answer questions 1 through 5.

They all drew to the fire, mother in the big chair with Beth at her feet, Meg and Amy perched on either arm of the chair, and Jo leaning on the back, where no one would see any sign of emotion if the letter should happen to be touching. Very few letters were written in those hard times that were not touching, especially those which father sent home. In this one little was said of the hardships endured, the dangers faced, or the homesickness conquered; it was a cheerful, hopeful letter, full of lively descriptions of camp life, marches, and military news; and only at the end did the writer's heart overflow with fatherly love and longing for the little girls at home.

"Give them all my dear love and a kiss. Tell them I think of them by day, pray for them by night, and find my best comfort in their affection at all times. A year seems very long to wait before I see them, but remind them that while we wait we may all work, so that these hard days need not be wasted. I know they will remember all I said to them, that they will be loving children to you, will do their duty faithfully, fight their bosom enemies bravely, and conquer themselves so beautifully that when I come back to them I may be fonder and prouder than ever of my little women."

Everybody sniffed when they came to that part; Jo wasn't ashamed of the great tear that dropped off the end of her nose, and Amy never minded the rumpling of her curls as she hid her face on her mother's shoulder and sobbed out. "I am a selfish girl! But I'll truly try to be better, so he mayn't be disappointed in me by and by."

"We all will!" cried Meg. "I think too much of my looks and hate to work, but won't any more if I can help it."

"I'll try and be what he loves to call me, 'a little woman,' and not be rough and wild; but do my duty here instead of wanting to be somewhere else," said Jo, thinking that keeping her temper at home was a much harder task than facing a rebel or two down South.

Beth said nothing, but wiped away her tears with the blue army sock and began to knit with all her might, losing no time in doing the duty that lay nearest her, while she resolved in her quiet little soul to be all that father hoped to find her when the year brought round the happy coming home.

Example:

What can we infer about Jo's personality?

A Jo has a good sense of humor.

B Jo is spoiled.

C Jo is proud and doesn't want to cry in public.

D Jo doesn't care about her father.

Answer:

C Jo is proud and doesn't want to cry in public.

1 Where can the reader infer that the father has gone?

A on a vacation
B to a camp in the mountains
C to war
D on a business trip

2 Based on the letter from their father, the girls know that:

A He wants them to maintain their good looks while he is gone.

B He wants them to work hard while he is gone.

C He wishes they were with him.

D He is very depressed by his surroundings.

3 Which of the girls might be described as a tomboy?

A Meg
B Beth
C Amy
D Jo

4 In this phrase, "the rumpling of her curls," what does the word <u>rumpling</u> mean?

A mussing
B styling
C combing
D covering

5 How long will their father be away from them?

A two years
B one year
C six months
D nine months

Directions: Read this passage from *Lives of the Hunted* by Ernest Thompson Seton and answer questions 6 through 11.

It was a rough, rock-built, squalid ranchhouse that I lived in, on the Currumpaw. The plaster of the walls was mud, the roof and walls were dry mud, the great river-flat around it was sandy mud, and the hills a mile away were piled-up mud, sculptured by frost and rain into the oddest of mud vagaries, with here and there a coping of lava to prevent the utter demolition of some necessary mud pinnacle by the indefatigable sculptors named.

The place seemed uninviting to a stranger from the lush and fertile prairies of Manitoba, but the more I saw of it the more it was revealed a paradise. For every cottonwood of the straggling belt that the river used to mark its doubtful course across the plain, and every dwarfed and spiny bush and weedy copse, was teeming with life. And every day and every night I made new friends, or learned new facts about the mudland denizens.

Man and the Birds are understood to possess the earth during the daylight, therefore the night has become the time for the four-footed ones to be about, and in order that I might set a sleepless watch on their movements I was careful each night before going to bed to sweep smooth the dust about the shanty and along the two pathways, one to the spring and one to the corral by way of the former corn-patch, still called the "garden."

Each morning I went out with all the feelings of a child meeting the Christmas postman, or of a fisherman hauling in his largest net, eager to know what there was for me.

Not a morning passed without a message from the beasts. Nearly every night a Skunk or two would come and gather up table-scraps, prying into all sorts of forbidden places in their search. Once or twice a Bobcat came. And one morning the faithful dust reported in grate (sic) detail how the Bobcat and the Skunk had

GO ⟹

differed. There was evidence, too, that the Bobcat quickly said (in Bobcat, of course), "I beg pardon, I mistook you for a rabbit, but will never again make such a mistake."

Example:

What can we infer from the above passage?

A   The author was wealthy.
B   The author was poor.
C   The author liked to have fun.
D   The author was well educated.

Answer:

B   The author was poor.

---

6   What is the Currumpaw?

A   a town
B   an animal
C   a friend
D   a river

---

7   How does the author describe his feelings in the morning when he goes outside?

A   He is sleepy and tired.

B   He is excited to see what he will find.

C   He is afraid he will run into an animal.

D   He is lonely and depressed.

---

8   What kind of message can we infer the Bobcat and Skunk leave for the author?

A   their fur
B   a note
C   their tracks
D   their sounds

---

9   In the last paragraph, the author says, "… the faithful dust reported in grate (sic) detail how the Bobcat and the Skunk had differed." What does this probably mean?

A   The animals had run around together all night.

B   The animals had shared a meal.

C   The animals had brought their young and compared their sizes.

D   The animals had fought during the night.

---

10   In the second paragraph, what is the best meaning for the word <u>denizens</u>?

A   inhabitants
B   people
C   plants
D   sculptures

---

11   The author states that the time for meeting new four-footed creatures is during the:

A   evening
B   day
C   night
D   winter

Directions: Read this passage from *The Wind in the Willows* by Kenneth Grahame and answer questions 12 through 17.

Baffled and full of despair, he wandered blindly down the platform where the train was standing, and tears trickled down each side of his nose. It was hard, he thought, to be within sight of safety and almost of home, and to be baulked by the want of a few wretched shillings and by the pettifogging mistrustfulness of paid officials. Very soon his escape would be discov-

GO

ered, the hunt would be up, he would be caught, reviled, loaded with chains, dragged back again to prison and bread-and-water and straw; his guards and penalties would be doubled; and O, what sarcastic remarks the girl would make! What was to be done? *He was not swift of foot;* his figure was unfortunately recognizable. Could he not squeeze under the seat of a carriage? He had seen this method adopted by schoolboys, when the journey-money provided by thoughtful parents had been diverted to other and better ends. As he pondered, he found himself opposite the engine, which was being oiled, wiped, and generally caressed by its affectionate driver, a *burly* man with an oil-can in one hand and a lump of cotton waste in the other.

"Hullo, mother!" said the engine-driver, "what's the trouble? You don't look particularly cheerful."

"O, sir!" said Toad, crying afresh, "I am a poor unhappy washerwoman, and I've lost all my money, and can't pay for a ticket, and I must get home tonight somehow, and whatever I am to do I don't know. O dear, O dear!"

"That's a bad business, indeed," said the engine-driver reflectively. "Lost your money—and can't get home—and got some kids, too, waiting for you, I dare say?"

"Any amount of 'em," sobbed Toad. "And they'll be hungry—and playing with matches—and upsetting lamps, the little innocents!—and quarreling, and going on generally. O dear, O dear!"

"Well, I'll tell you what I'll do," said the good engine-driver. "You're a washerwoman to your trade, says you. Very well, that's that. And I'm an engine-driver, as you may see, and there's no denying it's terribly dirty work. Uses up a power of shirts, it does, till missus is fair tired of washing 'em. If you'll wash a few shirts for me when you get home, and send 'em along, I'll give you a ride on my engine. It's against the Company's regulations, but we're not so very particular in these out-of-the-way parts."

The Toad's misery turned into rapture as he eagerly scrambled up into the cab of the engine. Of course, he had never washed a shirt

in his life, and couldn't if he tried and, anyhow, he wasn't going to begin; but he thought: "When I get safely home to Toad Hall, and have money again, and pockets to put it in, I will send the engine-driver enough to pay for quite a quantity of washing, and that will be the same thing, or better."

The guard waved his welcome flag, the engine-driver whistled in cheerful response, and the train moved out of the station. As the speed increased, and the Toad could see on either side of him real fields and trees, and hedges, and cows, and horses, all flying past him, and as he thought how every minute was bringing him nearer to Toad Hall, and sympathetic friends, and money to chink in his pocket, and a soft bed to sleep in, and good things to eat, and praise and admiration at the recital of his adventures and his surpassing cleverness, he began to skip up and down and shout and sing snatches of song, to the great astonishment of the engine-driver, who had come across washerwomen before, at long intervals, but never one at all like this.

Example:

What does the phrase "uses up a power of shirts" mean?

A Uses up lots of strong shirts
B Makes his shirts dirty
C Uses up two shirts
D Uses up lots of shirts

Answer:

D Uses up lots of shirts

---

12 Why is Toad not able to get a ride on the train?

A His enemies have chased him off.

B He has no money.

C He has a broken leg and can't get aboard.

D He has made the engine-driver angry.

13 The passage suggests that Toad has recently:

    A escaped from jail

    B sold his car and other belongings

    C lost his belongings

    D been very ill

14 In the first paragraph, the engine-driver is described as "burly." What is the meaning of this word?

    A slight in build

    B kindly appearing

    C muscular in build

    D tall

15 In the first paragraph, Toad says of himself, "he was not swift of foot." What does this phrase mean?

    A He is not a good dancer.

    B He has big feet.

    C He is not smart.

    D He cannot run fast.

16 Which of the following best describes what happens in this passage?

    A A washerwoman gets a ride so that she can get home to see her children.

    B Toad makes friends with the engine-driver.

    C Toad tricks the engine-driver so he can get a free ride home.

    D Toad agrees to become a washer-woman so the engine-driver will believe him.

17 In the last paragraph, the engine-driver is noted to be astonished and to have never seen a washerwoman like this before. What could you guess might happen after this?

    A Toad will learn how to wash shirts and do a good job.

    B Toad will arrive home without having any problems.

    C The engine-driver will grow to like Toad even more.

    D The engine-driver will discover that Toad is not a washerwoman.

STOP

LANGUAGE MECHANICS

Directions: Choose the answer with a punctuation error.

Example:

   A Have you seen
   B where Sally took the pliers?
   C I need them,"
   D Seth said.

Answer:

   A Have you seen

---

1 **A** "We are so proud of
     B the students who participate
     C in the spelling bee"
     D said Mr. Auld.

---

2 **A** What are the best ways to correct this.
     B It is hard to say, but
     C I'm sure there is a good
     D way to do it.

---

3 **A** We went to Disney World for New Year's Day.
     B What a wonderful trip it was.
     C We enjoyed every ride, show,
     D and display.

Directions: In the following examples, choose the sentence that is written correctly.

Example:

   A The book describes Ben Franklins adventures.
   B "What would you recommend?" she asked.
   C Please make dinner because Im hungry.
   D A lamp, a flashlight and a torch all shine.

Answer:

   B "What would you recommend?" she asked.

---

4 **A** We were driving to the mountains to spend a weekend at our favorite retreat.
     B The cabin is way back in the hills and it takes a long time to get there.
     C We enjoy swimming in the creek fishing in the river, and hiking in the fields.
     D After the weekend we are always tired but happy.

---

5 **A** In the summer, we like to go to the beach for at least a week?
     B We always save up our money plan to eat out at least once, and swim every chance we get.
     C Some people in our group enjoy playing miniature golf.
     D When I play golf I usually get at least one hole in one.

6 A Taking swimming lessons in the summer, is a lot of fun.

   B The instructors usually are fun-loving kind and talented

   C They like to work with children, and the children tend to like them back.

   D What do you think we can do after the lesson.

Directions: For questions 7 through 9, read the following paragraph and find the answer that shows the correct capitalization and punctuation.

yosemite national park (7) is a favorite vacation destination of many American families. Our family was there on july 4 1992 (8) and the park was very crowded. Most people advise that you should try to avoid the month of July if at all possible. By avoiding that month you (9) can probably find accommodations more easily and visit without having to deal with big crowds.

7 A Yosemite national park
   B Yosemite National park
   C Yosemite National Park
   D yosemite National Park

---

8 A July 4, 1992,
   B July 4, 1992
   C July 4 1992,
   D July 4 1992

---

9 A by avoiding that month you
   B By avoiding that Month, you
   C By avoiding that Month you
   D By avoiding that month, you

Directions: Read this story and use it to answer questions 10 through 12.

(10) hello said ellen as she entered the classroom.

The other girls were quiet until she came closer. She noticed that they were reading something in their laps. Ellen wondered what it was.
(11) "Hey, what's going on she asked.
"Oh, hi. We're reading about a contest coming up very soon. We really want to enter," Mallory commented.
"What kind of contest?" Ellen asked.
(12) "Oh, it's the annual speech contest sponsored by the rotary club," replied Mallory.

Directions: Choose the correct way to punctuate and capitalize the indicated phrases.

Example:

   A "where were you going?" sally asked.
   B "Where were you going? Sally asked.
   C "Where were you going?" Sally asked.
   D "Where were you going" Sally asked.

Answer:

   C "Where were you going?" Sally asked.

---

10 A "hello," said Ellen,
   B "Hello," said Ellen,
   C Hello," said Ellen.
   D Hello," said Ellen,

---

11 A "Hey, what's going on." She asked.
   B "hey, what's going on," she asked.
   C "Hey, what's going on?" she asked.
   D "Hey whats going on?" she asked.

---

12 A "Oh, it's the annual speech contest sponsored by the rotary club,"

   B "Oh, it's the annual speech contest sponsored by the Rotary Club,"

   C Oh, it's the annual speech contest sponsored by the Rotary club,"

   D "Oh its the annual speech contest sponsored by the Rotary Club,"

LANGUAGE EXPRESSION

Directions: Choose the word or phrase that best completes the sentence.

Example:

He was the _____ boy in his class.

A angrier
B angriest
C more angry
D most angrier

Answer:

B angriest

1 This is the _____ house on the street.

A quieter
B quietest
C most quiet
D more quiet

2 Annie _____ when it was time to go to bed.

A were working
B are working
C was working
D working

Directions: Choose the answer that is a complete and correctly written sentence.

Example:

A The water felt cold as we crossed the creek.

B Because Lia finished.

C Ellen captured the football and passes it.

D A winter day in the forest.

Answer:

A The water felt cold as we crossed the creek.

3 A We always like to go hiking when we goes to the mountain.

B They is our best friends in the world.

C When we go hiking, we always have a good time.

D When we go to the best restaurant in town.

4 A We builded a house on a hill so that we could see all around.

B Our house was pretty, although it was not very large.

C When we finished the house, it were our pride and joy.

D Packing up all of the stuff and working hard to get moved.

Directions: Read this story and use it to answer questions 5 through 7.

We tooked (5) a boat across the channel to explore the other side of the lake. We rided (6) for over an hour to reach the other side. Just as we were about to reach that side, the motor cut off, and we was (7) stuck. We asked the driver to find a paddle, but he couldn't find it. So we searched for one ourselves.

5 Which is the correct way to write "We tooked ..."?

A We took
B We taken
C We taked
D As it is written

GO

6 Which is the correct way to write "We rided …"?

A We roded
B We rode
C We ridden
D As it is written

7 Which is the correct way to write "we was stuck"?

A we were stuck
B we is
C we was sticked
D As it is written

Directions: For questions 8–9, mark the part of the sentence that is the simple subject.

Example:

The <u>horse</u> <u>has</u> a long <u>mane</u> which I <u>braid</u>.
     A **B**   **C**   **D**

Answer:

A horse

8 <u>Our</u> <u>teachers</u> are always very <u>involved</u> in
 A   **B**   **C**

<u>planning</u> the Spring Arts Festival.
 D

9 The <u>middle</u> <u>school</u> <u>has</u> a huge <u>courtyard</u>
     A   **B** **C**   **D**

that is used to display the artwork.

Directions: Mark the part of the sentence that is the simple predicate.

Example:

The <u>horse</u> <u>has</u> a long <u>mane</u> which I <u>braid</u>.
     A **B**   **C**   **D**

Answer:

B has

10 <u>Nicole</u> <u>is</u> a very good <u>artist</u> with several
   A **B**   **C**

<u>works</u> in the show.
 D

11 Carolyn <u>brings</u> some <u>pastels</u> and a <u>few</u> oil
        A   **B**   **C**

<u>paintings</u> as well.
 D

Directions: Read the paragraph below. Find the sentence that doesn't belong.

   Asian cooking is a fun hobby if you enjoy getting together with friends and cooking. (1) First, you have to find a good cookbook. (2) Then, you'll need to choose recipes. (3) Last year, we cooked a lot of Asian food in the summer. (4) You'll have to find a good Asian market that stocks the ingredients. Finally, you'll have to try out your skills with your friends around.

12 **A** Sentence 1
    B Sentence 2
    C Sentence 3
    D Sentence 4

Directions: Read the paragraph below. Find the sentence that best fits the blank in the paragraph.

   We also enjoy trying to find new Thai restaurants. One way to get recommendations is to talk to friends who have been to the restaurant you are considering. _____. Sometimes, however, you just have to try it and see for yourself.

GO ➡

13 **A**   I tend to eat Thai food in the winter.

  B   My friend Laurie especially likes Thai food.

  C   My favorite dish is a curry.

  D   You can also talk to members of the Asian community to see which restaurants they like.

STOP

SPELLING AND STUDY SKILLS

Directions: Find the correct spelling of the word.

Example:

The _____ of the school is very nice.

- **A** principle
- **B** principale
- **C** principal
- **D** princapal

Answer:

- **C** principal

---

1 Susan _____ wanted to have a good relationship.

- **A** sincerely
- **B** sincerly
- **C** sinserely
- **D** sinserly

---

2 The _____ candy was tempting to her as she tried to diet.

- **A** delishous
- **B** delicious
- **C** delishus
- **D** delicius

Directions: Choose the word that is spelled incorrectly.

Example:

- **A** occupy
- **B** oblige
- **C** rythmn
- **D** decision

Answer:

- **C** rythmn

3
- **A** lettuce
- **B** opportunity
- **C** rainey
- **D** afford

4
- **A** solar
- **B** scientists
- **C** atracts
- **D** technique

Directions: Choose the underlined word that is spelled incorrectly. If there is no mistake, choose No mistake.

Example:

We <u>dedicate</u> an hour <u>daily</u> to <u>practice</u>.
    A   **B**   **C**
<u>No mistake.</u>
    D

Answer:

- **D** No mistake

---

5 The <u>students</u> <u>huried</u> to their classes and
        A   **B**

<u>presented</u> their papers. <u>No mistake.</u>
    C   **D**

GO →

6 All of the <u>participants</u> had <u>there</u> notes and
             A   **B**

slides <u>ready.</u> <u>No mistake.</u>
     C   **D**

Directions: Look at the Table of Contents for a book below and answer questions 7 and 8.

Table of Contents
I.	Introduction to Rocks	Page 3
II.	Types of Rocks	Page 25
III.	Where to Find Rocks	Page 46
IV.	How to Start a Rock Collection	Page 58

Example:

In which chapter would you find the names of your rocks?

- **A** Chapter I
- **B** Chapter II
- **C** Chapter III
- **D** Chapter IV

Answer:

B Chapter II

---

7 In which chapter would you find information on how to arrange the rocks you have found?

- **A** Chapter I
- **B** Chapter II
- **C** Chapter III
- **D** Chapter IV

---

8 On what page would you find information about igneous rocks?

- **A** Page 3
- **B** Page 25
- **C** Page 46
- **D** Page 58

Directions: Shambra is writing a report about flowering shrubs. Answer questions 9 and 10 with that in mind.

9 Shambra used the book <u>Shrubs and Their Flowers</u> as one of her references. In what part of the book would she look to see references to azaleas?

- **A** the index
- **B** the glossary
- **C** the table of contents
- **D** the title page

---

10 What should Shambra include in her report?

- **A** the names of some ground covers
- **B** advice about when to plant flowering shrubs
- **C** names of neighbors who have flowering shrubs
- **D** information on trees in the area

---

11 Look at these guide words from a dictionary page.

honeybee	**hooves**

Which word could be found on this page?

- **A** home
- **B** horse
- **C** honor
- **D** honest

---

12 Which of these is a main heading that includes the other three words?

- **A** subways
- **B** trains
- **C** airplanes
- **D** transportation

MATH CONCEPTS

Directions: Choose the correct answer for the following questions:

Example:

Which digit means thousands in the numeral 8,546,901?

A 9
B 5
C 4
D 6

Answer:

D 6

1 Which digit means hundred thousands in the numeral 8,546,901?

A 9
B 5
C 4
D 6

2 Which of these is a prime number?

A 33
B 4
C 11
D 100

3 Which is another way to write 603,862?

A 600,000 + 30,000 + 800 + 60 + 2
B 600 + 300 + 80 + 60 + 2
C 6000 + 3000 + 800 + 60 + 2
D 600,000 + 3000 + 800 + 60 + 2

4 Where on the number line would $9^7/_{10}$ go?

A		B		C D
1 2 3 4 5 6 7 8 9 10 11 12				

A A
B B
C C
D D

5 Suppose Andrea was going to make Valentine baskets and wanted to put 50 pieces of candy in the baskets. How many baskets could she assemble that would contain exactly 7 pieces of candy?

A 8
B 6
C 7
D 9

6 Round 0.754 to the nearest tenth.

A 0.8
B 0.76
C 0.75
D 0.7

7 Which of these numerals would be read as "eight hundred ninety thousand, three hundred nineteen"?

A 800,090,319
B 890,319
C 89,319
D 8,009,319

GO

134

8 $\sqrt{81}$

   A 810
   B 8
   C 9
   D 8.5

9 What number is missing from this pattern?

$$\tfrac{1}{4}, .5, \underline{\quad}, 1.0$$

   A .6
   B .7
   C .8
   D .75

10 If you use only the digits 2, 5, 6, and 1, which of the following are the largest and smallest decimal numbers you can make?

   A .6251 and .1265
   B .6521 and .1256
   C .5621 and .2156
   D .6215 and .2165

11 The correct Roman numeral sequence is:

   A V, IV, IIV, X
   B IV, V, IVI, X
   C IV, V, VI, VII
   D IV, VI, VII, X

12 What is the correct number for the blank?

$$5 \times 76 = 5 \times (\underline{\quad} + 6)$$

   A 70
   B 60
   C 9
   D 10

STOP

MATH COMPUTATION

Directions: Choose the correct answer for the following questions.

Example:

97
×43

A 4181
B 4171
C 5171
D 6181

Answer:

B 4171

---

1 6.44 ÷ 46 = ____

A 1.4
B .14
C 14
D .114

---

2 378
×13

A 3914
B 4414
C 491
D 4914

---

3 $8\frac{3}{7} - 6\frac{5}{7}$ = ____

A $2\frac{5}{7}$
B $1\frac{5}{7}$
C $14\frac{8}{7}$
D $1\frac{1}{7}$

---

4 $78.31 + $46.99 = ____

A $125.30
B $126.39
C $115.30
D $125.20

---

5 0.8 × 54.8 = ____

A 438.4
B 4384
C .4384
D 43.84

---

6 8976
− 439

A 8533
B 8577
C 8537
D 7537

---

7 678
65
34
+ 23

A 800
B 801
C 8001
D 798

---

8 $6 \times \frac{5}{8}$ = ___

A 30
B $4\frac{1}{8}$
C $3\frac{3}{4}$
D $3\frac{3}{8}$

---

9 56.1 − 1.96 = ___

A 54.4
B 54.14
C 55.1
D 5.414

GO

10 $8^4/_5 - 5^3/_{10} =$ ___

    A $3^5/_{10}$
    B $3^3/_5$
    C $3^1/_5$
    D $3^3/_{10}$

---

11   11.601
    128.1
    +   .009

    A   139.71
    B   13.971
    C   1397.1
    D   139.819

---

12 $401 \overline{)8437}$

    A   21 R 26
    B   22 R 16
    C   21 R 16
    D   22

---

13   76
    $\times 43$

    A   3358
    B   3268
    C   3368
    D   4368

STOP

MATH APPLICATIONS

Directions: Choose the correct answer for questions 1–3.

1 What is the perimeter of this rectangle?

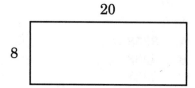

20

8

A 46 cm. **B** 160 cm. **C** 56 cm. **D** 28 cm.

2 What are the coordinates of point Q?

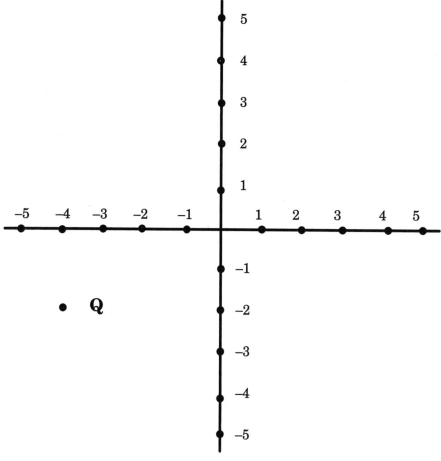

A (4, 2) **B** (−4, 2) **C** (−2, −4) **D** (−4, −2)

3 A seamstress has 125 yards of fabric. She uses 45 yards. Which number sentence tells how much fabric she has left?

   A $125 = 45 +$ ____
   B $125 + 45 =$ ____
   C $45 -$ ____ $= 125$
   D $125 +$ ____ $= 45$

Directions: Use the following chart to answer questions 4–6.

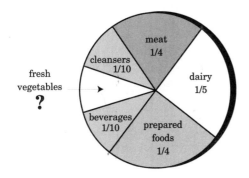

4 What fraction of grocery store purchases are in fresh vegetables?

   A $\frac{1}{5}$
   B $\frac{1}{4}$
   C $\frac{1}{10}$
   D $\frac{1}{3}$

5 What fraction of grocery store purchases are in meat and dairy products?

   A $\frac{1}{2}$
   B $\frac{9}{20}$
   C $\frac{2}{5}$
   D $\frac{8}{20}$

6 What is the value of y in this equation?

   $12 \times y = 132$

   A 23
   B 3
   C 11
   D 6

7 Which clock shows the time that is 3 hours and 15 minutes before noon?

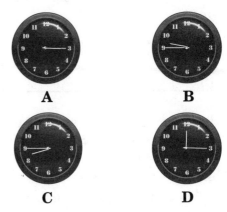

8 Which of the following containers would hold a package that required a volume of at least 120 cubic cm but not more than 140 cubic cm?

   A 4 cm × 6 cm × 4 cm
   B 5 cm × 5 cm × 5 cm
   C 5 cm × 6 cm × 10 cm
   D 4 cm × 5 cm × 3 cm

9 Jack, Julie, Hannah, and Dianne measure 62 inches, 59 inches, 54 inches, and 67 inches. What is the average height among these classmates?

   A 60
   B 59
   C 61.5
   D 60.5

10 Which of these shapes is a line segment?

   A ————————
   B ·————·
   C /
   D ∠

11 Nick has a pole that is 2.6 meters long. How many millimeters long is the pole?

   A   260

   B   26

   C   2600

   D   26,000

12 Heather pays for her jeans with $40.00 in cash. The jeans sold for $33.45. Which of the following is a possible combination of change she could receive?

   A   a five-dollar bill, a one-dollar bill, 1 quarter, and 1 nickel

   B   a five-dollar bill, a one-dollar bill, 1 quarter, 2 dimes, and 2 nickels

   C   a five-dollar bill, 2 quarters, and 1 nickel

   D   5 one-dollar bills, 4 quarters, 2 dimes, and 1 nickel

STOP

Answer Key for Sample Practice Test

Vocabulary

1	B
2	C
3	A
4	D
5	B
6	C
7	C
8	B
9	D
10	B
11	C
12	A
13	D
14	B
15	A
16	C
17	A
18	D

Reading Comprehension

1	C
2	B
3	D
4	A
5	B
6	D
7	B
8	C
9	D
10	A
11	C
12	B
13	A
14	C
15	D
16	C
17	D

Language Mechanics

1	C
2	A
3	B
4	A
5	C
6	C
7	C
8	A
9	D
10	B
11	C
12	B

Language Expression

1	B
2	C
3	C
4	B
5	A
6	B
7	A
8	B
9	B
10	B
11	A
12	D
13	D

Spelling and Study Skills

1	A
2	B
3	C
4	C
5	B
6	B
7	D
8	B
9	A
10	B
11	C
12	D

Math Concepts

1	B
2	C
3	D
4	D
5	C
6	A
7	B
8	C
9	D
10	B
11	C
12	A

Math Computation

1	B
2	D
3	B
4	A
5	D
6	C
7	A
8	C
9	B
10	A
11	A
12	C
13	B

Math Applications

1	C
2	D
3	A
4	C
5	B
6	C
7	C
8	B
9	D
10	B
11	C
12	B